THE FRONTIERS COLLECTION

AF862098

THE FRONTIERS COLLECTION

Series Editors:
A.C. Elitzur M.P. Silverman J. Tuszynski R. Vaas H.D. Zeh

The books in this collection are devoted to challenging and open problems at the forefront of modern science, including related philosophical debates. In contrast to typical research monographs, however, they strive to present their topics in a manner accessible also to scientifically literate non-specialists wishing to gain insight into the deeper implications and fascinating questions involved. Taken as a whole, the series reflects the need for a fundamental and interdisciplinary approach to modern science. Furthermore, it is intended to encourage active scientists in all areas to ponder over important and perhaps controversial issues beyond their own speciality. Extending from quantum physics and relativity to entropy, consciousness and complex systems – the Frontiers Collection will inspire readers to push back the frontiers of their own knowledge.

Information and Its Role in Nature
By J. G. Roederer

Relativity and the Nature of Spacetime
By V. Petkov

Quo Vadis Quantum Mechanics?
Edited by A. C. Elitzur, S. Dolev, N. Kolenda

Life – As a Matter of Fat
The Emerging Science of Lipidomics
By O. G. Mouritsen

Quantum–Classical Analogies
By D. Dragoman and M. Dragoman

Knowledge and the World
Challenges Beyond the Science Wars
Edited by M. Carrier, J. Roggenhofer, G. Küppers, P. Blanchard

Quantum–Classical Correspondence
By A. O. Bolivar

Mind, Matter and Quantum Mechanics
By H. Stapp

Quantum Mechanics and Gravity
By M. Sachs

Extreme Events in Nature and Society
Edited by S. Albeverio, V. Jentsch, H. Kantz

The Thermodynamic Machinery of Life
By M. Kurzynski

The Emerging Physics of Consciousness
Edited by J. A. Tuszynski

Weak Links
Stabilizers of Complex Systems from Proteins to Social Networks
By P. Csermely

Mind, Matter and the Implicate Order
By P.T.I. Pylkkänen

Paavo T.I. Pylkkänen

MIND, MATTER AND THE IMPLICATE ORDER

With 8 Figures

Springer

Paavo T.I. Pylkkänen
University of Skövde
Consciousness Studies Programme
School of Humanities and Informatics
Box 408
541 28 Skövde, Sweden
e-mail: paavo.pylkkanen@his.se

Series Editors:

Avshalom C. Elitzur
Bar-Ilan University,
Unit of Interdisciplinary Studies,
52900 Ramat-Gan, Israel
email: avshalom.elitzur@weizmann.ac.il

Mark P. Silverman
Department of Physics, Trinity College,
Hartford, CT 06106, USA
email: mark.silverman@trincoll.edu

Jack Tuszynski
University of Alberta,
Department of Physics, Edmonton, AB,
T6G 2J1, Canada
email: jtus@phys.ualberta.ca

Rüdiger Vaas
University of Gießen,
Center for Philosophy and Foundations of Science
35394 Gießen, Germany
email: Ruediger.Vaas@t-online.de

H. Dieter Zeh
University of Heidelberg,
Institute of Theoretical Physics,
Philosophenweg 19,
69120 Heidelberg, Germany
email: zeh@urz.uni-heidelberg.de

Figure credits: Cindy Tavernise: page 62; Matthew Zimet: page 163; Chris Philippidis, Chris Dewdney and Basil Hiley: pages 167 and 168; David Bohm: pages 172 and 200
Cover figure: Image courtesy of the Scientific Computing and Imaging Institute, University of Utah (www.sci.utah.edu).

ISSN 1612-3018
ISBN-13 978-3-642-06286-5 e-ISBN-13 978-3-540-48058-7

This work is subject to copyright. All rights are reserved, whether the whole or part of the material is concerned, specifically the rights of translation, reprinting, reuse of illustrations, recitation, broadcasting, reproduction on microfilm or in any other way, and storage in data banks. Duplication of this publication or parts thereof is permitted only under the provisions of the German Copyright Law of September 9, 1965, in its current version, and permission for use must always be obtained from Springer. Violations are liable for prosecution under the German Copyright Law.

Springer is a part of Springer Science+Business Media

springer.com

© Springer-Verlag Berlin Heidelberg 2007
Softcover reprint of the hardcover 1st edition 2007

The use of general descriptive names, registered names, trademarks, etc. in this publication does not imply, even in the absence of a specific statement, that such names are exempt from the relevant protective laws and regulations and therefore free for general use.

Cover design: KünkelLopka, Werbeagentur GmbH, Heidelberg

To Elina, Kasper, and Ester, my implicate, explicate orders

Preface

I have been planning to write a book ever since I finished my PhD dissertation *Mind, Matter and Active Information: The Relevance of David Bohm's Interpretation of Quantum Theory to Cognitive Science*, in 1992. For various reasons it has taken me a long time to do this. I started writing notes in February 2000. The project became more concrete when in March 2003 I talked with Jack Tuczynski and the opportunity arose to offer a book for inclusion in Springer's new *Frontiers Collection.* It has now been more than two years since the more concrete plan for the book emerged. That time has not just been spent "writing up" ideas that I already had. It has additionally meant rethinking many of the problems and ideas. I have also read a great deal, including re-reading most of Bohm's philosophical contributions – 25 years after I first read some of them. Rethinking and re-reading is exciting but also potentially frustrating because it often makes you realize how complex and fascinating the problems are in themselves and how far we actually are from solving them!

This book deals with topics that have been variously neglected and even "forbidden" in academic circles during much of the 20th century. One such issue is conscious experience, the study of which was famously suppressed in behaviorist psychology and even in traditional cognitive science, and has only relatively recently become the focus of intense research in a number of fields.

Another neglected and suppressed area is metaphysics, often defined as the study of the most basic and general features of reality and our place in it. While metaphysics had over many centuries – indeed millennia – been a central area of philosophy, there was a strongly anti-metaphysical attitude in much of 20th-century philosophy, in such dominant philosophical trends as logical positivism and ordinary-language philosophy, while in, say, Heideggerian phenomenology there was a strong urge to renew traditional metaphysics. There has, however, been a comeback of metaphysics during the past decades, especially within the so-called analytical philosophical tradition.

Yet another traditionally "forbidden" topic to be discussed in this book connects with physics and has to do with the interpretation of quantum theory, the theory that deals with atomic phenomena. To some extent, the suppression of discussion about the meaning of quantum theory is connected with the above-mentioned suppression of metaphysics in general, for quantum theory was initially interpreted in the 1920s in the strongly anti-metaphysical climate of logical positivism. As a result, much emphasis was placed on how the theory predicts the results of experiments, while it was thought to be meaningless to speculate about the nature of quantum reality over and above the experimental phenomena. Now, the history of science shows that it has frequently been fruitful to make hypotheses about currently unobservable phenomena; but those who tried to do this in the early days of quantum physics met with, at times, dogmatic opposition. Whereas philosophers fairly soon began to realize the limits of positivism, it exerted a strong influence upon quantum physics until fairly recently, making it "forbidden" to try to sketch the nature of reality at the quantum level.

Thus, during the 20th century, speculating about the nature of conscious experience, the nature of matter, and the fundamental nature of reality was often off limits. One might wonder what it was that one was allowed to speculate about! What this book tries to do is to study the interrelations of these traditionally forbidden topics. My first concern is with how modern quantum and relativity physics, and theories that build upon them, might influence our more general concept of reality; I then consider whether conscious experience and its relationship with matter could be understood in a new way in the context of such a new "quantum metaphysics". Of course, it is no longer forbidden to study consciousness, or the nature of reality, or the meaning of quantum theory, although raising these issues might still set some eyebrows in motion. However, it is still relatively rare to consider these topics together. One potential advantage is that doing this might give rise to a new "big picture", which in turn might have a profound effect upon our more detailed efforts. That is why I am concerned with the big picture.

Is there really a need for such a synthetic approach?, one might wonder. To see why it is necessary, consider, for example, metaphysics. Professional philosophers are again busy exploring such traditional metaphysical issues as substance, identity, universals and particulars, modalities (possibility, necessity, contingency), and so on. However, much of this research does not take into account developments in modern natural science, especially in physics. But surely the natural sciences can contribute to a better understanding of the topics central to metaphysics? In turn, I think that an understanding of traditional metaphysics, and philosophy more generally, can be of great help when trying to make sense of the puzzling features of, say, quantum phenomena. The idea is that philosophy and physics can mutually constrain, criticize, and inspire each other as we move toward a better theory.

How about the mind and conscious experience, then? Is it really necessary to consider metaphysics or physics when theorizing about them? Isn't it enough to focus on those subjects which more directly and obviously study the mind, such as psychology, philosophy of mind, cognitive science, cognitive neuroscience, artificial intelligence, and so on? The key point here is that the philosophical problem of consciousness in our materialistic age is typically raised by first taking it for granted that there is a physical world and then asking whether we can locate or find a place for consciousness in such a physical world. But what does "physical" mean? Clearly, contemporary physics can at least prima facie have something important to say about this. It seems to me that many of the "mind sciences" presuppose a late 19th-century view of the physical world and formulate the philosophical problems of consciousness in relation to such a view. But it is plainly part of the scientific attitude to revise our basic notions, including our notion of the "physical", if experimental and theoretical developments call for this. Thus physics may after all play a key role when we are trying to understand the place of consciousness in nature.

Of course, the approach adopted in this book is very ambitious. Areas such as metaphysics, quantum theory, and consciousness are already difficult when considered separately. When such difficult areas are considered together, there is obviously a risk of confusion. One thus needs to proceed very carefully and rely on expertise in the different domains. It is also important to pay attention to previous similar efforts and their successes and shortcomings. This is partly why I have chosen in this book to focus so strongly upon David Bohm's views. For it seems to me that Bohm was one of the few 20th-century thinkers who had a good grasp of not only quantum physics (which was his original field of expertise and which he indeed helped to develop) but also the natural sciences more generally, as well as philosophy and consciousness. He saw the importance of trying to understand the connections between these areas and was indeed developing a new "big picture", based on his research in quantum and relativity physics, as well as on his own philosophical explorations. In my view, Bohm's efforts deserve a much more careful study than they have hitherto been given.

Given the vastness of the Bohmian programme, this book is necessarily only a small step toward a better understanding and evaluation. At the very least, I hope that it will help others to judge better whether this programme provides fruitful tools for tackling the fascinating problems concerning matter and consciousness that are at issue. I believe that scientific and philosophical theories are primarily ways of looking, tools that help us to see, to understand. In the course of writing and reading and discussing when preparing this book, I certainly have had many exciting moments. We are, in a sense, prisoners of time, but there is nothing like a new notion of time to help to see our existence from another viewpoint. An unexpected sense of freedom can come with such a shift of perspective.

Acknowledgements

This book, like so many other things in life, both good and bad, was initiated in a bar over a glass of cold beer. It was April 2003 in Tucson, Arizona, and the second "Quantum Mind" conference, excellently organized by Stuart Hameroff, had just finished. Some of us were at the lobby bar of Hotel Congress. I chatted with the physicist Jack Tuszynski, whom I had got to know through Stuart and who is one of the leading experts on the biophysics of microtubules. I was enroute to Stanford University for a couple of months' research visit, with an ambition to start writing a book. Jack had recently become editor of Springer's *Frontiers Collection*. There was a book project in search of a publisher, and an editor in search of books. I am very grateful to Jack for encouraging me to submit this book for consideration to the *Frontiers Collection* and would also like to give my warmest thanks to the editorial staff at Springer, especially Dr. Angela Lahee and my copy editor Mark Seymour.

I am also extremely grateful to Stuart for his tremendous help and friendship over the years. He has been a major force behind the emerging field of "consciousness studies", having a key role in organizing the biannual Tucson "Toward a Science of Consciousness" conferences since 1994 and many other related activities, not least those dealing with the relevance of quantum theory to consciousness studies. In other words, he has done much to create the very conditions and atmosphere that facilitate the sort of research this book tries to do. Thanks for all of this!

One of the few philosophers attending the "Quantum Mind" conference in 2003 was David Chalmers. I had gotten to know Dave in connection with the earlier Tucson conferences, and he had even visited us at the University of Skövde, Sweden, in 2000. His work on consciousness has inspired me and many others, and I am grateful for the many discussions with him over the years. When Dave heard that I had the possibility of going to Stanford, he suggested that I contact Dr. Edward Zalta, the director of the Metaphysics Lab at Stanford University. Ed was able to arrange for me Visiting Scholar status at the Department of Philosophy at Stanford University and a place to work in the famous Center for the Study of Language and Information (CSLI), where the Metaphysics Lab is located. We had a number of stimulating discussions with Ed, and he further introduced me to other researchers and visitors in the Center. All this was very inspiring, and I was able to get this book project off the ground. Needless to say, I am extremely grateful for having had the opportunity to visit Stanford University. Thanks, Ed, for your hospitability, and thanks, Dave, for mediating the contact!

Some words of personal history might be useful here to illuminate the more long-term background of the book, as well as my indebtedness to another Dave, namely the physicist-*cum*-philosopher David Bohm, whose work is the focus of this book. I became interested in Bohm's work around 1979 when I had begun my studies in philosophy at the University of Uppsala, Sweden. I first met him in 1980 at Brockwood Park Educational Centre in Hampshire, England, and got to know him through the various discussions he was leading in the Centre during the early eighties. We started a closer cooperation while I was doing my Masters degree in Logic and Scientific Method at the University of Sussex, Brighton, in 1984–5. This led to a workshop – initiated and co-organized with Dr. Juan Hancke – in Sussex in 1986 and an anthology, *The Search for Meaning*, focusing on Dave's philosophical ideas, which I edited and which was published in 1989.

In the late 1980s I was back in my country of origin, Finland, studying toward a PhD in Theoretical Philosophy at the University of Helsinki. Funded by the Academy of Finland, I was, however, able to spend time in the USA (Southern California) and especially in the UK (Oxford and London), which enabled me to have many discussions with Dave, who was Professor of Theoretical Physics at Birkbeck College, University of London, and also visited California regularly.

We had been having discussions together for a number of years, and began to work on a joint paper, tentatively entitled "Cognition as a movement towards coherence" in the spring of 1990 in London. By 1991, Dave felt that we ought to develop the article into a book. In October 1992 he had almost finished the book he was working on with his long-time colleague Basil Hiley, *Undivided Universe*, and I had completed my PhD dissertation earlier that year. Dave suggested that now was the time to continue our work. A week later, while I was preparing to leave for the UK, I got a telephone call telling me that Dave had died of a heart attack.

David Bohm's death was a tremendous loss for all who knew him. I lost a teacher and a colleague but above all a close and dear friend. Our joint project was left unfinished, although I hope one day to be able to edit it, perhaps supplemented with some transcripts of recorded discussions between us, into a publication. Some glimpses of our discussions are even provided in the present book at the end of Chapter 3.

Dave's widow, Saral Bohm, has also been a good and supportive friend over the years, for which I am very grateful. She has been playing a key role in many of the activities Dave was engaged in, such as the experiments with the dialogue process.

Hopefully this background will help the reader to further understand why I have been focusing so sharply on Bohm's work. To some extent, I am just trying to push a little further forward the project we were engaged in and for which, of course, Dave's thinking provided the main framework and background. At the same time, of course, it is not possible to continue a joint project alone. Thus this book is a mixture of a study of Bohm's ideas (in order to understand them better) and an attempt to use these ideas and even go beyond them in tackling the difficult problems that are at issue.

Another person who has been extremely important and helpful for the development of my ideas and the writing of this book is Professor Basil Hiley, also at Birkbeck College, University of London. Dave introduced me to Basil in 1991, and our contact soon grew into active cooperation and friendship. Basil generously arranged for me to be a Research Fellow at the Department of Physics at Birkbeck during 1994–6, and we have met regularly since then. I have learnt tremendously from him, although, of course, he is not to be held responsible for any mistakes in my understanding of physics! Basil and I have written a number of joint articles, mostly on the mind–matter relationship in the light of Bohm and Hiley's interpretation of quantum theory. These papers – as well as many of Basil's own papers on the topic –

are somewhat more technical than the discussion in the present book. They provide an important complement to the present discussion.

Basil read parts of the manuscript for this book and made many valuable comments. Most importantly, he has encouraged me over the years to keep on working on these issues, and his thinking has provided a constant source of inspiration. I hope that the present book will help others to understand better the philosophical motivation behind Basil's recent research, which takes the Bohmian programme into new domains.

There are, of course, many other important long-term influences on my thinking, and here it is possible to mention just a few of them. An important period was in 1990–95, when I was an Academy of Finland researcher based in Helsinki, in a project directed in an inspiring way by Dr. Antti Hautamäki. My colleague in the project, Dr. Pauli Pylkkö, played a particularly important role in deepening my understanding of both analytical and continental philosophy. I think his 1998 book *The Aconceptual Mind*, although relatively little known, is one of the most ground-breaking philosophical studies that have appeared in recent years. He is introducing new philosophical themes and giving direction to some old debates in an original way. Needless to say, I feel lucky to have had the opportunity to learn from him and work closely with him. His influence can also be seen from time to time in the present book. One of the researchers who has realized Pylkkö's importance is Professor Gordon Globus of the University of California, Irvine. Gordon has also been a constant support for my research, and indeed has through his own innovative work paved the way for the sort of synthesis of science and philosophy that I, too, am striving towards.

Another important influence is Dr. Antti Revonsuo, of Turku University, Finland and now a guest professor at the University of Skövde. Antti, too, was part of the Academy of Finland research programme in cognitive science and is now an internationally well-known researcher in the field of consciousness studies. Antti and I have shared an interest in consciousness since the early 1990s, and perhaps simply because we have different perspectives we have had many interesting and intense discussions. I certainly have learnt much from these discussions. Furthermore, Antti's critical comments prompted me to choose the phenomenal structure of consciousness, and time consciousness in particular, as one focus of my research. I have for many years had the intuition that quantum theory is relevant to the understanding of consciousness and its place in nature, but have had difficulties pinning down exactly what that relevance is. Antti's criticism was that quantum models rarely address conscious experience directly, i.e. its phenomenal and structural properties. This led me to consider Bohm's notion of the implicate order more carefully, for I realized that Bohm had proposed that the implicate order prevails in phenomenal experience itself (rather than just at the quantum level). To some extent, the present book is an attempt to answer Antti's valuable criticisms of the so-called "quantum approaches to consciousness".

I would also like to mention Dr. Ron Chrisley of Sussex University, England. Ron has been a good friend and critic over many years. He has in particular helped me to understand the connection between some Bohmian notions of meaning and contemporary debates on the philosophy of language and representation.

Special thanks belong to the Finnish philosopher Dr. Erkka Maula, who gave me invaluable guidance, especially during my early attempts to enter the worlds of philosophy and science.

I would also like to thank those who have specifically influenced or commented on the manuscript. First of all, I am much indebted to Anna-Karin Selberg, who is a PhD student in philosophy at the University of Skövde and the University of Södertörn, Stockholm. Chapter 5 of the present book, the one on time consciousness, was written in the context of an intense and fruitful discussion with Anna-Karin, and she presented some of the results of our discussion at the Second Annual Meeting of the Nordic Network for Consciousness Studies in Turku, Finland in May 2003. She has also challenged me to look at philosophical themes from a Heideggerian perspective, which at times has had a strong effect upon my thinking and being.

I would like to thank Dr. Tere Vadén, a philosopher now based in Tampere University, Finland. Tere was for many years my closest colleague at the University of Skövde. He has helped my research in countless ways and has also made valuable comments on parts of the manuscript. He is, in my view, also one of the few contemporary philosophers who are able to move across the boundaries between continental and analytical philosophy, and between philosophy and natural science, extending to cognitive science, media studies, and more.

I have been a "Docent" or a kind of Adjunct Professor at the Department of Philosophy, University of Helsinki, since 2001. In practice, this means that I regularly do some teaching there. This connection has turned out to be very valuable for the present book, as it has enabled me to have contact with the young generation of Finnish philosophers and cognitive scientists, and in particular two of its foremost representatives: Otto Lappi and Anna-Mari Rusanen. They helped me to get my writing going when I was stuck. In addition, Anna-Mari has read a fair amount of the manuscript and made very valuable comments, often leading me to revise sections considerably. It has been great to get that kind of support from one of my home towns, Helsinki, where much of my research began. I am also pleased that one of my Helsinki students, Ari-Pekka Lappi, has recently completed an interesting Masters thesis on Bohm's notion of the rheomode of language, partly under my supervision.

Yet another Helsinki-based researcher who has commented on the manuscript is Dr. Tarja Kallio-Tamminen. Tarja is engaged in a similar project, i.e. seeking to unify quantum theory, metaphysics, and consciousness. While I am focusing on the Bohmian scheme, Tarja has focused on Niels Bohr's scheme. Her comments have often helped me to see my work from another perspective.

I have been Associate Professor in Theoretical Philosophy at the University of Skövde, Sweden since 1996. This has enabled the sort of very long-term basic research that is necessary for the present kind of book, which tries to tackle the big picture. I am grateful to the various people in Skövde who have understood the potential value of this kind of research and made it possible for me to develop my ideas in this way, in particular the former Vice Chancellor Lars-Erik Johansson, as well as my supportive Heads of Department Seppo Luoma-Keturi, Ulf Wilhelmsson, and Johan Almer.

One of the specialities in Skövde is the presumably first "consciousness studies" undergraduate programme in the world, combining philosophy and cognitive neuroscience with conscious experience as a focus. The programme, which I have helped to create, has provided a framework for my teaching and research in Skövde. I am honored to have been able to work and talk with Stefan Berglund, Monica Bergman, Per Hansell, Sakari Kallio, and Anders Milton, among others. Special thanks belong also to my students and especially to my Masters-level philosophy students in Skövde. I began to lecture on Bohm's notion of the implicate order to my Masters-level course in the autumn of 2001, and this has considerably helped me to understand the various difficult notions. Some of my students, such as Peter Ericson, Roque Molina, Karl Nilsson, and Elisabeth Trenter, have gone on to write interesting Bachelors and Masters theses connected with Bohmian themes. I am also grateful to Khashayar Naderehvandi and Hans Sjöberg for their inspiration.

I should also like to thank my other Swedish colleagues. At the University of Gothenburg, the linguists Jens Allwood and Elisabeth Ahlsen in particular helped me to get my research on the implicate order off the ground when I lived in Gothenburg in 1998–2000. In Stockholm, where I have lived since 2000, I have particularly enjoyed contact with Agora for Biosystems, under the auspices of the Swedish Royal Academy of the Sciences. The core group of Agora, Hans Liljenström, Peter Århem, B. Ingemar, and B. Lindahl have provided an excellent framework for thinking and discussion, and supported my research in many ways. This has also facilitated contact with Professor Ingemar Ernberg's philosophy of biology group at the Karolinska Institute, Stockholm. The group's seminars have been very stimulating and have helped me to keep track of the latest developments in biology.

There are a number of other people, discussions with whom have been very helpful over the years, including Harald Atmanspacher, Srinivas Aravamudan, Uzi Awret, Bernie Baars, Matti Bergström, Frank Borg, Don Factor, Laurence I. Gould, Irmeli and Antti Hautamäki, Scott Hagan, Ivan Havel, Nick Hobbs, Allan Hobson, Ted Honderich, Janne Karimäki, Synnöve Karvinen-Niinikoski, Ranjana Khanna, Andrei Khrennikov, Timo Laiho, Jim Laukes, Petteri Limnell, Katherine McGovern, India Morrison, Ilkka Niiniluoto, David Peat, Barbara Piechocinska, Arkady Plotnitsky, Veikko Rantala, Andreas Roepstorff, John Searle, Jouko Seppänen, Maxim Stamenov, Stig Stenholm, Kyriakos Theodoridis, Sampo Vesterinen, Giuseppe Vitiello, Benkt Wankler, Georg Wikman, Nancy Woolf, and Tom Ziemke. Of course, they are not to blame for the shortcomings of this book! Those not mentioned are not forgotten.

I would also like to acknowledge the support given by both my family and my wife Elina's family, as well as by our many friends.

No book project will suffer from having to spend some time in Paris. As my wife Elina, a national economist, got an offer to work in the OECD in Paris during 2004, it didn't take us too long to accept. I remember in particular the long walks in the Bois de Boulogne, where we often discussed various aspects of the implicate order. Indeed, whether in the Yosemite mountains, Big Sur, the Latin Quarters of Paris, the norra Mälarstranden in Stockholm, or Palojoki, Huittinen, Finland, it is during the many walks and discussions with her that much of the understanding required to write this book has been worked out. I am more than grateful to her for that, as well as for her love and friendship.

Södra Ängby, Stockholm *Paavo Pylkkänen*
May 2006

Contents

1

Introduction

1.1 Preamble

In this book we shall be considering some questions that have a long history. These include questions about the fundamental nature of matter and its movement; the nature of mind and its relationship to matter; and the nature of time, both physical and mental. We will also be concerned with how these questions are connected with one another. For example, what relevance might our theories about matter have to our views about the relationship between mind and matter?

Of course, we are in a very different position to tackle these questions today than were those who first formulated them. When it comes to the nature of matter and physical time, we now have advanced theories in physics; when it comes to the nature of mind, there are likewise advanced theories both about the phenomenal (e.g. spatio-temporal) structure of the mind, as well as about how mental processes correlate with the underlying physical processes in the brain and the body.

But regardless of these advances, there are many aspects of these questions that remain unclear. In physics, the basic theories are quantum theory and relativity. It is well-known, however, that the interpretation of quantum theory has been the subject of intense debate ever since the theory was first developed in the 1920s. Quantum theory predicts the results of experiments (e.g. in the atomic domain) with brilliant accuracy. Mathematically, one uses the Schrödinger equation and its solution, the famous wave function, to accomplish the predictions. But how should we interpret the wave function? Is it merely a mathematical *tool*, a part of an algorithm for predicting the probability of finding a particle (e.g. an electron) in a given small region in a measurement (as Bohr thought)? Or is the wave function a complete *description* of the electron, which is thought to be wave-like when it moves but which collapses into a particle in a measurement in which we always observe the electron as a particle (as von Neumann thought)? Or is the wave function a complete description of the electron, which is thought to be wave-like when it moves, while the appearance of a particle in a measurement is explained not by the collapse of the wave but rather by assuming that the universe *branches* into "many worlds", each with a particle in a different position (as Everett and de Witt suggested)? Or is the wave function a description of just one part of the electron, namely a field aspect that *guides* another part of the electron, namely a particle aspect, so that there is no need to assume a collapse of the wave or the branching of the universe to explain why we observe a particle (as de Broglie and Bohm thought)? These questions are still actively debated, and it is fair to say that the meaning of non-relativistic quantum theory remains unclear and that the jury is still out, perhaps even a bit "far out" at times.

Furthermore, although both relativity and quantum theory work brilliantly in their own domains, their basic concepts seem to be in complete contradiction with each other. Thus, as Bohm has underlined, relativity emphasizes continuity, locality, and determinism, while quantum theory suggests that the exact opposite is fundamental, namely discontinuity, non-locality, and indeterminism. This strongly suggests the need for a yet broader and deeper theory in physics, containing relativity and quantum theory as limiting cases that work in their own domains. There are proposals for such a new theory (e.g. string theory and loop quantum gravity theory), but these remain fairly speculative (see Weinstein (2006)). Thus, although everyone agrees that the classical Newtonian and Maxwellian notion of matter is completely wrong in certain domains, and that quantum theory and relativity are required to deal with many known physical phenomena, there is not yet agreement about what the more fundamental theory of matter is that can unite relativity and quantum theory and describe all known physical phenomena in a coherent and unified way.

Physics, from a philosophical point of view, is also characterized by a great deal of conceptual confusion. For example, it is customary to talk about "elementary particles", evoking the image of there being some absolute, fundamental building blocks or tiny "billiard balls" that interact mechanically with each other, and out of which the mechanical "clockwork universe", including bodies and brains as its parts, is constituted. However, it has been known since the 1920s that such "particles", besides having particle properties (such as mass, charge, and momentum), also exhibit wave-like properties (diffraction and interference) and even properties that strongly violate any mechanistic scheme (non-local correlations and discontinuity of movement). The concept of "elementary particles", and the images it may evoke, is thus actually very limited in its ability to help us capture what is essential about what we might call the more *fundamental architecture* of the physical world, as revealed in quantum and relativistic phenomena. We need new concepts and images that can better illuminate features such as wave–particle duality, non-locality, and the discontinuity of movement. But there is not yet agreement as to what such concepts should be, and consequently, a great deal of confusion prevails in attempts to discuss the more fundamental structure of the physical world.

This state of affairs also has consequences for other subjects. Thus, for example, in philosophy there has been a strong tendency to look to the natural sciences when trying to resolve traditional philosophical issues, a tendency known as "physicalism" (see, for example, Stoljar (2001)). However, when one examines the work of the leading physicalists, one can see in them little *systematic* effect from, say, quantum and relativity physics, or later developments in physics. To be sure, there is some effect (e.g. emphasis upon the relativistic notion of an event (Davidson 2001)), but on the whole, physicalism remains a relatively empty research programme, instead of relying upon some specific proposal about the nature of physical existence that would do justice to contemporary physics. In a nutshell, physicalism says that our general concept of reality ought to be some sort of a generalization of what the natural sciences, especially physics, tell us (see, for example, Quine (1960) and Koskinen (2004)). But as a matter of fact, most physicalist views currently on offer seem to have a very weak relationship to modern quantum and relativity theory. Physicalism thus does not yet manage to do what it says it ought to do.

This "hollowness of contemporary physicalism" creates a great deal of frustration in philosophy. There are difficulties in the very attempt to formulate problems, let alone in the various attempts to solve them. For when physicalists formulate a philosophical problem, they typically make a reference to the physical world. Questions that are formulated and debated today include: What is the relationship between mental phenomena and the physical processes in the brain and matter more generally? What is the relationship between meaning and the physical items that carry meaning? However, as long as there is no coherent notion of what the physical means, the very problems making a reference to the physical will be out of focus (cf. Montero (1999)). Typically, physicalist philosophers rely upon some common-sense notions about the physical world that more or less resemble the ideas of 19th-century classical physics. But this is, of course, in violation of the stated aim of the physicalist programme, namely that philosophy should rely upon the best natural sciences rather than upon, say, common sense or theories that have shown to be very limited.

Of course, classical physics is a brilliant achievement that still works approximately correctly in a wide range of domains. But it gives *completely wrong* predictions about centrally important domains of the physical world. It typically fails in the domain of very small distances and very small energies, where quantum theory is needed. But there are also macroscopic quantum effects, visible to the unaided eye, such as superconductivity, superfluidity, and Bose–Einstein condensation. Furthermore, classical physics also fails to account for such more familiar macroscopic properties as the stability of matter, the temperature of the Sun, and bulk specific heats. It is thus certainly a mistake to think that quantum theory is irrelevant to explaining the properties of the physical world as we encounter them in everyday experience. On the contrary, one could argue, most of these properties can be explained in terms of the quantum theory. To give yet another example, the wavelengths of the light emitted by atoms can only be understood in terms of quantum theory, thus implicating quantum theory in the physical understanding of colour, a familiar, everyday property of the world.

All this suggests a challenge for modern philosophy. On the one hand, many philosophers are tempted by physicalism, saying, for example, that our general concept of reality (or ontology) ought to be some sort of generalization of what the natural sciences, especially physics, say. On the other hand, it has turned out to be very difficult to take into account what physics, in particular, has to say. What is urgently needed, therefore, is some reasonably general, intelligible account of the results of modern physics, if the Emperor of Physicalism is to ever to put on some clothes.

There are also well-known difficulties in attempts to understand the nature of mind and its relationship to matter. Mind and matter seem very different in their basic qualities and yet they seem intimately related, so much so that many have tried to reduce mind to matter, suggesting that mental processes are *identical with* some neurophysiological processes in the brain. However, such reductive attempts have been questioned. Many philosophers have suggested that *conscious experience* presents a particularly serious problem to mind–body reductionism, because it has many features that seem very different from objective, neurophysiological processes. These include the qualitative character of conscious states (the "raw feels" or "qualia"; for example, the taste of a strawberry milkshake); their subjectivity (e.g. only I seem to have direct access to my inner conscious states, such as my experience of pain); and their meaningfulness or "intentionality" (e.g. my conscious states typically have meaning to me, but how can anything mean anything to anything in a purely physical system?).[1] The greatest puzzle has to do with the simple fact that when I am conscious there seems to be something we might call "experiencing" going on. But what is such "experiencing" and how does it arise? How could objective, physical processes give rise to "experiencing", which at least *seems* to be something altogether different from objective physical processes?

It seems obvious that whatever else we may be, we definitely are experiencing beings. It seems equally obvious that "experiencing" is not something independent of physical processes, but rather is closely correlated with them. Just think, for example, of the dream–wake cycle. There is a part of our sleep when we are not conscious, but when we are dreaming we are conscious and while we are awake we are conscious. There are neural correlates of the dream–wake cycle, suggesting that the brain is strongly implicated in "experiencing".

So it seems obvious that "experiencing" is *correlated with* neural processes, but not at all obvious that it is *nothing but* neural processes. In fact, it seems obvious (at least to me) that experiencing cannot be identical with the sorts of mechanical neurophysiological processes that modern neuroscience talks about.

According to modern neuroscience, consciousness, of course, has to do with your brain and nervous system, and the body more generally. So it is typically assumed by neuroscientists that a physical system, made of certain components that interact, is conscious. To put it very simply, there are nerve cells in the brain organized in particular ways to make anatomical regions and connected with each other in complex ways, transmitting information through electrical action potentials, but also in more subtle ways such as chemical pathways.

Most current neural theories of consciousness are expressed in terms of the activities and connections of the neurons. Thus, for example, there is the idea of consciousness having to do with re-entrant connections between brain regions; the idea that consciousness is essentially connected with thalamo-cortical loops; the idea that consciousness has to do with synchronized "40 Hz" oscillations of electrical activity in

[1] For a very good recent overview of the problems connected with consciousness, see van Gulick (2004). See also Chalmers (1996).

the brain; and the idea that consciousness essentially involves a certain "global workspace" implemented in the brain. (For various neural theories of consciousness see, for example, Baars et al. (2003).) But the question is, why should such mechanical interactions between physical parts make you conscious? I say "mechanical", because most neural theories of consciousness, from the point of view of physics, only appeal to the level of the classical physics of Newton, Maxwell, and the like. And classical physics is mechanical. It has to do with particles moving along trajectories under the influence of forces (gravitational and electromagnetic) and colliding with each other, and with fields (i.e. waves in the electromagnetic field) that influence charged particles and are influenced by them, a bit like the way a water wave can mechanically set a rubber duck in motion, while moving the duck in calm water will produce waves.

It is common in neuroscience to think that when it comes to physics, only neural processes that obey classical physics are required to explain consciousness. A typical idea is that we need a large network of such mechanically behaving neurons to give rise to consciousness (e.g. tens of thousands of neurons). But how could such purely mechanical activity of particles and fields in your brain, not violating the laws of classical physics, give rise to consciousness? Let us construct a simple thought experiment to explore this. Let us say that I am given mechanical components that are structurally equivalent to all the components that the modern neural theories appeal to (e.g. suitable artificial neurons), and I set them up so that the system as a whole is functionally equivalent to the functions that modern neural theories appeal to (e.g. there are re-entrant connections, thalamo-cortical loops, 40 Hz synchronized oscillation, a global workspace, etc. in my system of artificial neurons). This might be difficult in practice but surely conceivable in principle. Will the artificial system be conscious? It seems obvious to me, and to many others, that it will not. It seems that conscious experiencing is something that cannot be derived from mechanical physical processes. But if so, what is it then?

In fact, Leibniz had already realized this difficulty, as has been succinctly described recently by Robert van Gulick:

> In the *Monadology* (1720) [Leibniz]...offered his famous analogy of the mill to express his belief that consciousness could not arise from mere matter. He asked his reader to imagine someone walking through an expanded brain as one would walk through a mill and observing all its mechanical operations, which for Leib-

> niz exhausted its physical nature. Nowhere, he asserts, would such an observer see any conscious thoughts. (van Gulick 2004)

A similar problem is still actively debated in contemporary philosophy of mind, perhaps today best known under the label "the hard problem of consciousness" (Chalmers 1995, 1996). Of course, not everyone agrees that consciousness is such a hard problem. Consider, for example, Daniel Dennett:

> Might it be that somehow the *organization* of all the parts which work one upon another yields consciousness as an emergent product? And if so, why couldn't we hope to understand it, once we had developed the right concepts? This is the avenue that has been enthusiastically and fruitfully explored during the last quarter century under the twin banners of cognitive science and functionalism – the extrapolation of *mechanistic naturalism* from the body to the mind. After all, we have now achieved excellent mechanistic explanations of metabolism, growth, self-repair, and reproduction, which not so long ago also looked too marvellous for words. Consciousness, on this optimistic view, is indeed a wonderful thing, but not *that* wonderful – not too wonderful to be explained using the same concepts and perspectives that have worked elsewhere in biology. Consciousness, from this perspective, is a relatively recent fruit of the evolutionary algorithms that have given the planet such phenomena as immune systems, flight, and sight. (Dennett 1999)

Thus, if Dennett is correct, consciousness might be "an emergent product" in a mechanical system, provided the parts are organized in a suitable way. But as long as we are not given any clue about how "experiencing" could arise from the interactions of parts, the reference to "emergence" is no better than the Cartesian reference to God as the source of consciousness and the mediator of the interaction between matter and consciousness. Reference to "emergence" surely sounds these days scientifically more respectable than reference to God. But is it really any more enlightening? *How* does conscious experience emerge from the mechanical neurophysiological processes in the brain?

One possibility, advocated by Dennett, is that the same mechanical concepts and perspectives that have worked elsewhere in biology will also work for consciousness. Another possibility is that they will not, or not beyond a certain point. In physics, we have seen that concepts and methods that worked well for a given domain of physical phenomena (i.e. the classical domain) fail completely in a wider domain (i.e. the relativistic and quantum domains). Perhaps the same will turn out to be the case in biology. The mechanistic concepts and methods that work well for metabolism, growth, etc. may fail for some important aspects of consciousness. In biology and psychology, just as in physics, we may then need radically new theories. One of the key ideas to be explored in this book is that the new theories in physics may actually help us to develop the sorts of new theories in biology and psychology that may be required to give an adequate explanation of consciousness. This, as such, would be nothing new. The theories of physics have often influenced the theories of mind. Often, however, this has given rise to overly mechanical theories of mind. But perhaps the more holistic theories of contemporary physics will help to inspire theories of mind that can better do justice to the holistic features of the mind.

Of course, one may ask whether it is necessary or even desirable for theories of physics to affect theories of mind at all. One reason why such influence may be inevitable is that physics deals with general categories such as space, time, movement, and causality, which are relevant to almost everything, and certainly to mind. Physics helps us to understand many of our most general concepts better and suggests changes in them; and once you change your general concepts, you will see the world in a new way.

Let us move on to briefly consider some further problems concerning the relationship between mind and matter. One such problem that has been vigorously debated in recent philosophy of mind has to do with the *causal powers* of the mind. Mind seems to be very different from matter (because of some of the features of conscious experience such as subjectivity, inner qualitative feels, meaningfulness, the very fact that experiencing is going on, etc.), but it also seems obvious that our mental states – both conscious and unconscious – influence the behavior of our body. But how are we to make sense of this influence? For example, if minds are not described by the laws of physics, should the laws of physics be modified to allow for the causal influence of minds upon bodily behavior? One of the aims of this book is to provide new ways of thinking about this problem, known as the *problem of mental causation* (see, for example, Robb and Heil (2005)).

There are also problems connected with the spatio-temporal structure of conscious experience. We have already mentioned some features of consciousness that are difficult to relate to matter, such as qualia, subjectivity, and meaningfulness or "intentionality". Yet another important aspect of consciousness is what we might call the *phenomenal structure* of conscious experience. Although the terms "qualia" and "phenomenal properties" are sometimes used interchangeably in the literature, it is useful to distinguish phenomenal structure from the qualitative structure of conscious experience. This has been recently emphasized and succinctly described by van Gulick:

> "Phenomenal organization" covers all the various kinds of order and structure found within the domain of experience, i.e., within the domain of the world as it *appears* to us. There are obviously important links between the phenomenal and the qualitative. Indeed qualia might be best understood as properties of phenomenal or experienced objects, but there is in fact far more to the phenomenal than raw feels. As Kant (1787), Husserl (1913), and generations of phenomenologists have shown, the phenomenal structure of experience is richly intentional and involves not only sensory ideas and qualities but complex representations of time, space, cause, body, self, world and the organized structure of lived reality in all its conceptual and nonconceptual forms. (van Gulick 2004)

There are some paradoxical features associated with phenomenal organization. For example, our experiences typically have a temporal structure, perhaps most evident in situations such as listening to music. However, when listening to music, we are not merely apprehending a process that proceeds step by step, say, paying attention to one note/chord now and another a bit later. No, in a musical experience we also seem to perceive a melody as a whole, a theme that grows, develops, and transforms. Typically, of course, we do hear some notes for the first time "now", but we also seem to directly perceive (rather than, for example, just passively remember) the notes that were first heard some time ago, and also anticipate the perception of future notes. We perceive a *whole structure* that is in some sense "timeless". Yet the usual view of time says that only the present and what is in it exists. But if this is true, how can we then, for example, when listening to music *perceive* (as opposed to just remember) a structure that includes the notes heard a little time ago, which latter, according to the usual view of time, no longer exist? Husserl thought that we perceive the past but admitted that this is like saying there is "wooden iron". This paradox of "time consciousness" is one of the issues that has been debated both in traditional phenomenology and contemporary philosophy of mind and cognitive science (see, for example, Dainton (2000), van Gelder (1999), and Varela (1999)). Another of the aims of this book is to explore a new way of looking at time consciousness.

We have seen in our brief overview that on the one hand, there are difficulties in developing a coherent notion of matter, and on the other hand, there are difficulties in understanding the nature of mind and its relationship with matter. There are also puzzling questions about the relationship between the structure of conscious experience and our usual notion of time. There is something worth noting at this point. When philosophers, psychologists, cognitive scientists, cognitive neuroscientists, etc. consider problems like the mind–matter problem, the problem of mental causation, and the problem of time consciousness, they usually consider some key concepts such as "matter", "causality", "movement", and "time" – implicitly or explicitly – in the spirit of the classical physics of Newton, Maxwell, and others. Of course, we have already referred to this above when we noted that contemporary physicalist philosophers typically ignore quantum and relativity physics, or that neuroscientists typically think that the neural correlates or constituents of consciousness are mechanical. Now, it *could* be the case that those material processes which play a relevant role in the mind–matter relationship and/or our phenomenal experience all lie in the domain of classical physics, that is, in the domain where classical physics provides a good approximation. However, no-one has been able to show *how* the whole of the mind can be reduced to such classically conceived matter. In particular, in the case of conscious experience, we could say that no one has come *anywhere near* to showing this. This is what Leibniz' analogy of the mill points to, and this, in my view, is what David Chalmers' well-known work suggests.

In fact, one could argue that a large proportion of the problems concerning mind and matter are problems that arise in relation to the classical notion of matter. If such problems could be solved, then perhaps philosophers discussing the mind–matter problem could safely ignore non-classical physics. But given the fact that no-one has been able to solve the mind–matter problem for the matter of classical physics, one reasonable possibility is to explore whether, in order to understand the relationship between mind and matter better, we need to consider matter in the light of our broader and more accurate theories, such as relativity and quantum theory. For we *know* (in so far as we know anything at all in science) that classical physics gives wrong predictions in some domains of the physical world, and it thus cannot be considered an adequate theory of the whole of matter known to us at present. Could it be that some of the physical processes that enter centrally into the relation of mind and matter could lie outside of the domain of classical physics? I am not claiming that they do so, but given the failure of contemporary mind sciences to relate central aspects of the mind to the classical domain, I suggest that this is an option worth considering.

One might be tempted to see physics in terms of its domains by saying that classical physics describes a certain domain, say A, quantum physics describes another domain B, and relativity theory yet another domain C, etc. One might further be tempted to assume that such domains are separate and independent of each other, and that the physical world is made up of such domains. I think there are good reasons to question such a way of thinking. Firstly, there are various kinds of relationships between the domains – for example, it is the stability of atoms that quantum laws establish which enables, say, the table to exist as a relatively solid macroscopic object. Furthermore, I think it is also interesting to consider the view that a given theory of physics suggests something about the general architecture of the physical world. Thus, Newtonian physics fits well with an atomistic architecture and the idea of a universe as a huge machine. However, quantum theory and relativity theory suggest that although the universe has a mechanistic sub-domain, some other architecture is more fundamental. For example, instead of emphasizing that the universe is made up of its parts, these theories might emphasize the primacy of the whole, and see the parts as derivative.

So I think that physics is concerned not just with separate levels of nature, but also with the general architecture of nature. This gives rise to one important way in which physics can be relevant to the understanding of the mind. For one of the traditional philosophical problems concerning the mind is not only the relationship between mind and matter, but also the broader, more architectural question about the place of mind in nature (see Broad (1925) and Chalmers (2002b)). Quantum theory and relativity theory strongly suggest that we need a new concept of the general architecture of physical reality, and clearly this at least may be relevant when trying to locate the mind and conscious experience in the physical world. I thus think it is not at all clear that quantum theory and relativity can be safely ignored by the "mind sciences" on the basis that they deal with strange and different domains of the physical world, although many researchers seem to think so.

If we agree with the above line of thinking, our challenge is to explore whether the relationship between mind and matter could be understood better if our notion of matter were based on post-classical physics, such as quantum theory or relativity, or some even better theories inspired by them. Here, however, we immediately run into the problems already mentioned above. There is much disagreement about the interpretation of quantum theory, and there are serious problems in trying to relate quantum theory and relativity to each other. There are well-known attempts to develop new, more fundamental theories (e.g. string theory and loop quantum gravity theory), but these attempts are very speculative and still far from being satisfactory. Thus, even if we wanted to try to relate mind to some post-classical notion of matter, this is difficult simply because it is not clear what a coherent post-classical notion of matter is!

To sum up, then, it seems fairly certain that we cannot satisfactorily solve the mind–matter problem, when matter is understood in terms of classical physics. But we also know (insofar as we know anything at all in science) that classical physics, although approximately correct in a certain domain, is completely wrong in other domains. This naturally gives rise to the possibility that perhaps mind and matter connect with each other in the domain that lies outside that of classical physics. In any case, mind could find its place in nature better if our notion of the general architecture of nature was inspired by modern physics. Consequently, the relationship between mind and matter might be understood better if our notion of matter were based on post-classical physics. But there we run into the problem that post-classical physics does not yet provide us with a commonly accepted, coherent new notion of matter.

Now, a very interesting attempt to try to tackle *both* the question about the nature of matter in the light of quantum theory and relativity *and* the question about the nature of mind and its relation to matter was carried out by the physicist-*cum*-philosopher David Bohm (1917–1992). In his work we find, among other things, a new proposal about the more fundamental architecture of matter; a proposal about the nature of the mind and how it relates to matter; and even a proposal about how to make sense of time consciousness (for example, how we might be able to perceive "past" elements of experience, which the usual view of time says do not even exist). Might Bohm's work help us to go forward in tackling the difficult questions that we have briefly described above? One of the main aims of this book is to explore this question. Let me therefore proceed to give a brief overview of Bohm's ideas on matter, mind, and their relationship, including his ideas about time and conscious experience. After this, to conclude this introductory chapter, I will briefly explain what the rest of the book tries to do.

1.2 Bohm on Matter, Mind, and Their Relationship

David Bohm was educated as a physicist and made some significant contributions to mainstream physics, working on plasma, metals, and liquid helium. For example, Bohm and Pines' plasma theory of electrons in metals was the first theory to coherently explain the stability of metals (Pines 1987). However, Bohm became more and more interested in philosophical questions in his research. Are electrons waves or particles? Are quantum processes genuinely indeterministic? Are there quantum jumps – that is, is movement discontinuous at the quantum level? Are there instantaneous, "non-local" correlations between spatially separated particles, and does this create problems with the theory of relativity? More generally, is it possible to have a single coherent model of systems at the quantum level, or are we forced to be satisfied with "complementary" but mutually exclusive modes of description (such as wave and particle), as Niels Bohr had famously emphasized? Bohm's research tackles these questions from many different perspectives and suggests different answers, depending on the perspective chosen. Perhaps one of his greatest achievements was precisely to show the possibility of a number of different perspectives in quantum theory. We still do not know which perspective is the correct one; but in the meantime, it is useful to know what the coherent options are, and Bohm certainly made a significant contribution to developing and clarifying the options.[2]

An important influence on Bohm's philosophical development was Einstein, with whom he had many discussions while in Princeton in the late 1940s and early 1950s. Bohm had also early on become interested in Niels Bohr's philosophically sophisticated interpretation of quantum theory. When one looks at his scientific and philosophical contributions as a whole, it would not be completely inaccurate to place him somewhere between Einstein and Bohr. For example, with Einstein, Bohm shared the view that the task of physics is to try tell us something about a reality that exists independently of ourselves. With Bohr, he shared the view that quantum theory emphasizes undivided wholeness,

---

[2] Bohm had a dramatic life at times, including political problems in the USA during the McCarthy era in the early 1950s. David Peat's (1996) biography *Infinite Potential: The Life and Times of David Bohm* provides a vivid account of his life. Many researchers have suggested that some of Bohm's ideas about quantum theory were simply suppressed rather than evaluated in the spirit of open, fair, and genuine criticism. For various sociological studies connected with Bohm, see, for example, Beller (1999), Cushing (1994), Freire Jr. (2005), Olwell (1999), and Pinch (1977). See also Forman (1987).

as well as the more philosophical idea that it is important to carefully consider the role of language and communication in physics.

Bohm's first major philosophical contribution was, perhaps surprisingly, his quantum mechanics textbook *Quantum Theory*, published in 1951. This book, which explicates quantum theory from the conventional Copenhagen (i.e. Bohr's) point of view clearly shows the importance Bohm gave to a more philosophical understanding of quantum theory, over and above focusing on mathematical and technical aspects, which was beginning to be the dominant trend in physics. *Quantum Theory* is also important for our theme of mind and matter, for it contains a section in which Bohm discusses striking analogies between quantum processes and the process of thought (see Bohm (1951, pp. 168–172) and Pylkkänen (2004b)). However, this thorough attempt to explicate quantum theory under a Bohrian interpretation left Bohm dissatisfied. Discussions with Einstein – who was a well-known critic of Bohr's interpretation of quantum theory – further prompted him to look for another interpretation of quantum theory. Einstein and Bohm were particularly dissatisfied with the extremely empiricist, "positivistic" feature of the usual quantum theory, which did not allow one to discuss reality beyond the observations. Observations, in turn, were fairly limited (e.g. a spot appearing in a photographic plate), so it seemed that quantum theory was providing a truncated, fragmented view of reality. Einstein himself had been seeking a more realist and causal interpretation of quantum theory as early as the 1920s, but without success.

Bohm came up with two different ideas. One of them introduces the idea of an "incoming wave" to account for a quantum mechanical measurement. In standard quantum theory, a particle such as an electron is mathematically described by a wave function. The wave typically spreads out over a large region, but whenever we measure the electron we always find it as a small particle-like entity in a very small region of space (e.g. making a spot on a photographic plate). There is thus a contradiction between the mathematical wave description (which, when taken as a description of the electron, suggests that the electron is a wave that is typically spread out) and the particle description (which we use to describe what we actually observe in every measurement of an individual electron). Different ways of resolving this contradiction give rise to different interpretations of quantum theory. Some said that the wave function should not be taken as a description of the electron, but instead should be seen as a "probability wave", a mathematical tool that we can use to calculate probabilities of finding the electron, conceived of as a little particle, in a certain small region in a measurement. This "minimal" interpretation is consistent, but it goes strongly against the intuition that physics ought to provide a description of individual systems. Thus others went on to suggest that the wave function does describe the electron. But how then do we deal with the above contradiction? The key new assumption was to say that this wave collapses into a small region in a measurement, thus giving rise to the particle-like manifestation we actually observe. This approach is better than the minimal one because it provides a description of quantum processes even before measurements. However, the notion of the collapse is problematic. It has proved difficult to give it a coherent description, thus making it seem too much like an ad hoc solution to the problem.

This is where Bohm's idea of an "incoming wave" comes in. In a typical experiment, the electron is described in terms of a wave that is spreading out, and yet we always find the electron as a particle. Bohm thought that perhaps the reason we see a particle is not that the outgoing wave suddenly collapses, but that there is another wave closing in to that point – either the original, outgoing wave somehow reflected back, or else a new wave. Symmetry has often played an important role in physics, and there is certainly a quality of symmetry in this proposal by Bohm. If there are outgoing waves, why not incoming ones? What we call a "particle" could then be seen as a certain phase in the movement of these waves, namely that phase when the incoming wave has closed in, giving rise to an intense, particle-like pulse in a small region. Bohm did not, at the time, further pursue this idea of an electron being an aspect of a process of outgoing and incoming waves. However, as we will see later, it plays an important role in his "implicate order" approach, with its notions of "enfoldment" and "unfoldment", which he began to develop in the 1960s.

The second idea Bohm had after his discussions with Einstein soon gave rise to a more concrete proposal and publications (see Bohm (1952)). Remember again that the basic problem of standard quantum theory is that the mathematics we use to describe the electron suggests that the electron is a wave, while every measurement reveals the electron as a tiny particle. So the electron seems to be sometimes a wave and sometimes a particle, making it a very ambiguous entity. But what if the electron is *both* a wave *and* a particle *all the time*? Its having a wave aspect would then explain why it obeys the mathematics of wave behavior; while its particle aspect would explain why we always find a particle when we measure the electron. Note in particular that if we assume that the electron is always both a particle and a wave, there is no need to assume that the wave collapses in order for us to explain why we see a particle. The reason why we observe a particle is simply that the electron always is a particle. And the reason why we observe such particles obeying the mathematics of wave behavior (e.g. in the famous two-slit experiment) is simply that each particle has a wave associated with it and guiding it.

Bohm indeed proposed, independently discovering and further developing an idea de Broglie had already in the 1920s, that we should look at an individual quantum system, such as an electron, as a combination of a particle and a new kind of wave, described by the wave function. He actually arrived at the idea when considering the relationship between quantum theory and classical physics. Strictly speaking, both quantum theory and classical physics apply to the same reality. The usual idea is that quantum theory is the more accurate, general, and fundamental theory. Classical physics can then be seen as a special limiting case that can be derived from quantum theory and that works approximately well in certain domains. However, things were not that simple in standard quantum theory. For as we saw, according to the usual interpretations, the quantum theoretical description of reality is either minimal or ambiguous. Either there is no view of quantum reality beyond measurements at all, or else quantum reality consists of systems that are ambiguously sometimes waves, sometimes particles. If classical physics is supposed to be a special case of quantum theory, then presumably the everyday reality that classical physics describes ought to be somehow derivable from the reality that quantum theory describes. But how could you derive the solid everyday reality from no reality at all or from an ambiguous reality? Bohm's insight was to look at the equations of quantum theory and to see that there actually was a well-defined view of quantum reality implicit in them, a quantum reality from which one could derive the everyday classical reality in a coherent way. To put it very schematically, Bohm saw that quantum reality consists of systems that have a particle aspect and a wave aspect. The wave aspect influences the particle aspect, giving rise to all the strange quantum phenomena we observe in experiments (e.g. electron interference, non-locality). However, whenever the new quantum mechanical wave aspect has a negligibly small effect, we are left with just the particles obeying the laws of classical physics and – bingo! – we have derived classical physics from quantum physics.

This is one of the beautiful aspects of Bohm's "ontological interpretation": it shows us how the classical, familiar everyday world arises from the more exotic quantum world under certain circumstances. Another issue that the ontological interpretation raises is the question of causality at the quantum level. The usual interpretation of quantum theory had suggested that individual quantum processes are indeterministic. On the other hand, in the ontological interpretation, the wave aspect of the electron guides the particle aspect, suggesting that the behavior of individual quantum processes might be deterministic after all. Bohm felt that the situation called for a very careful, more philosophical study of the question of causality and chance, and this resulted in his book *Causality and Chance in Modern Physics*, which was published in 1957. His basic proposal was that both causality and chance are always needed whenever we are dealing with some limited domain of the physical world. Thus, for example, he was not claiming that the quantum level was completely deterministic. Instead, the determinism suggested by the ontological interpretation ought to be seen as a statistical average of chance fluctuations at a deeper level. A closer study of these chance fluctuations might, in turn, reveal some more lawful behavior, which might, however, turn out to be a statistical average of a yet deeper level of chance fluctuations, and so on. Bohm felt that there was no need to assume a fundamental level, and thus the question whether the fundamental level is deterministic or indeterministic would not even arise (see Bohm (1957, 1986)).[3]

The ontological interpretation provides a very useful perspective to quantum phenomena. However, it is also a limited perspective. The challenge for modern physics is to unite quantum theory and the theory of relativity, in particular quantum theory and general relativity. To do this, it is necessary to go deeper than the ontological interpretation by itself can take us. Thus, in the 1960s, Bohm, together with his colleague Basil Hiley, began to develop a more general framework for physics in which one could hope to be able to unite quantum theory and relativity. This more general framework he later called the "implicate

[3] Bohm's 1952 papers in *Physical Review*, which are the basis of the ontological interpretation of quantum theory (Bohm & Hiley 1993), were originally characterized by him as dealing with "hidden variables" in the quantum theory. The approach has also been called the "causal interpretation of quantum theory" and the "pilot wave theory". However, Bohm felt in the end that the essential point of the interpretation is that it makes a hypothesis about the nature of quantum reality, and not so much its deterministic features or the idea that the variables are hidden. These 1952 papers have given rise to a number of approaches; see, for example, Albert (1992, 1994), Bedard (1999), Bell (1987), Cushing et al. (1994), Goldstein (2002), Holland (1993), and Valentini (2001). In this book I am partly focusing upon Bohm's own further development of the original 1952 approach, developed in particular with Basil Hiley and their research students at Birkbeck College, University of London.

order" framework. He also extended this framework to biological and psychological phenomena, proposing it as a more general metaphysical theory of reality as a whole. Finally, in the late 1980s and early 1990s, we see him trying to bring together his two main schemes – that is, the more specific ontological interpretation of quantum theory and the more general implicate order framework. He thought that the ontological interpretation can help to extend and specify the implicate order framework both as a theory of physics and as a theory of the relationship between mind and matter.

It is clear that Bohm was concerned with providing a description of reality – at the quantum level, and more generally, a unified description of matter, life, and consciousness, all adding up to a general concept of reality or a metaphysical theory. However, it is important to realize that although he was clearly more concerned with describing a mind-independent reality than many other 20th-century physicists or philosophers, this concern did not mean that he ignored the role of the mind (language, perception, etc.) in his attempts to describe reality. In other words, he did not ignore epistemological issues or questions that concern the nature of our knowledge and the problems of justifying it. On the contrary, his broad philosophical work includes extensive studies of various epistemic issues: physics and perception (Bohm 1965a), the notions of truth and understanding (Bohm 1964), a view of science as "perception-communication" (Bohm 1977), experimentation with the structure of language (Bohm 1977), study of knowledge understood as process (Bohm 1974), and discussions of topics such as communication, creativity, art, and so on. To fully appreciate Bohm's views about the nature of reality, they should be understood in the context of his epistemic considerations. Although our focus in this book will be upon ontological questions related to matter and consciousness, this should not be taken as a sign that I consider the epistemic issues unimportant. The reason I am focusing here upon the ontological issues is partly that ontological issues have often been ignored in recent science and philosophy, and partly that a proper consideration of the epistemic issues would make this book too large.

Let us proceed to consider Bohm's general ontological views head on, focusing on his views about the relationship between mind and matter. The strategy will be first to describe the notion of implicate order as it applies to matter; then to consider how it applies to mind; then to note that the implicate order framework needs to be extended in order to provide a better view of the relationship between mind and matter; then to consider Bohm's notion of "soma-significance" as one such extension;

and finally to consider how the ontological interpretation of quantum theory can be used to extend the implicate order framework to provide a better mind–matter theory.

Many people, including myself, have found Bohm's idea of the implicate order difficult to grasp. Indeed, I remember from discussions with Bohm how he was keen to emphasize that the idea was at a fairly early stage of development. Let us therefore begin to unpack it by considering a very succinct description he provided in a 1990 article, "A New Theory of the Relation of Mind and Matter" (Bohm 1990). The basic idea of the implicate order is that

> the whole universe is in some way enfolded in everything and that each thing is enfolded in the whole. This implies that in some way, and to some degree, everything enfolds or implicates everything. However, this takes place in such a manner that under typical conditions of ordinary experience, there is a great deal of relative independence of things. (Bohm 1990, p. 273)

Such an idea of "enfoldment" of the whole universe in each part, which resonates with Leibniz's idea of monads and William Blake's poetry, may seem very counterintuitive, exotic, and strange at first. But as we will see later, enfoldment is taking place in a wide range of domains, and actually right there in front of you. Think of the small region of space where your eye is placed. In this region, there is a movement of electromagnetic waves (light waves) that carries the information you use as the basis for constructing your visual experience. This movement somehow contains or "enfolds" information about the whole room, or if you happen to be watching the night sky, about the whole universe of space and time. This enfolded information is then *unfolded* by the lens of your eye, and later in a very complex process by your brain, resulting, when combined with information supplied by your brain, in your visual experience of a three-dimensional world with objects in it. Of course, as already mentioned above, we do not really understand *how* the objective physiological process becomes a conscious visual experience, but we shall discuss that problem later.

Note further that such enfoldment of information about the whole into each small region typically takes place in *all* wave phenomena, for example in sound waves. Thus, when you go to a concert to listen to a symphony orchestra, information about what each instrument plays is typically enfolded in each region, including the one where your ear is placed. But you have to be quiet, because if you were to speak loudly, information about what you say will likewise be momentarily enfolded in the movement of air molecules in every region, and others might not enjoy your contribution to the enfolded order!

Now, according to quantum field theory, which physicists widely consider the most accurate theory of matter currently on offer, even elementary particles are understood in terms of an underlying activity of fields. Thus, according to this theory an "elementary particle" such as an electron is not just a little billiard ball, although some aspects of its behavior suggest that it has particle-like features (for example, in a measurement, the electron always appears in a very small region of space, and it has measurable particle properties such as mass, charge, and spin). More fundamentally, however, it is thought that the electron is based on the activity of a field, which is in some ways similar to the activity of light waves. The "electron field" gives rise, at least momentarily, to a particle-like manifestation, when there is an intense field in a small region, or a localized pulse. Bohm pointed out that the mathematics of quantum field theory suggests that similar enfoldment and unfoldment that is found with light waves and sound waves also prevails in the movement of quantum fields that underlie all matter. Thus, just as light waves in a small region can enfold information about the whole universe, so the waves that underlie each "elementary particle" can similarly enfold information about the whole universe.

There is thus a sense in which each region or "part" of the universe enfolds information about the whole universe. But we can look at this also from another point of view. With light, we can say that, typically, information about a part can be found in every single region, throughout the whole of space. Consider, for example, the book you are holding. Information about the book is enfolded in each region of the room, in the movement of the light waves. Or imagine that you go outside when there is a full moon. Information about the Moon is enfolded in every region of space where the light reflected from the Moon has travelled. So, more generally, we can say that not only is information about the whole enfolded in each part, but information about each part is also enfolded in every region of the whole.

According to Bohm's interpretation of quantum field theory, this also applies to "elementary particles". Underlying each such particle is a movement of a field. This movement enfolds information about the whole universe into the small region where the field manifests itself as a particle-like entity. But because the field is also spread, in principle, throughout the universe, information about the particle-like entity can be found in every region of the universe. In this sense, the whole universe is enfolded in everything, and everything is enfolded everywhere in the whole universe. The implicate order thus prevails as the most fundamental order of the universe currently known to us.

Of course, this is a very exotic idea. Just think of all the atoms and particles that constitute your body. We are used to thinking about them as tiny little things that just passively sit there. But quantum field theory, as interpreted by Bohm, suggests otherwise. There is a sense in which each particle in your body enfolds information about the whole universe (analogously to the way the activity of light waves in the region where your eye happens to be placed can enfold information about the whole universe). There is also a sense in which information about each particle in your body is enfolded throughout the universe (analogously to the way information about a planet is enfolded in every region of the space in which there are light waves reflected from the planet). The proposal is that, as a part of the universe, each one of us thus enfolds information about the whole universe, not only via our senses, especially vision, but also, and more exotically, via the underlying field nature of the very "particles" that constitute our body. The further suggestion is that through various movements of fields (light and, most fundamentally, quantum fields) information about us is enfolded throughout the whole universe.

Just think how different this proposal is from the traditional mechanistic view of matter and the physical world that was developed by Galileo, Newton, and others! It is common, in this world view, to think of the Earth and human beings as a mere speck of dust in a huge cosmos, externally related to other things, and governed by completely mechanical laws. In contrast, Bohm's interpretation of quantum field theory suggests that there are not just external, but also, and more fundamentally, internal, relationships between the part and the whole and, via the relation to the whole, between the parts themselves. Obviously, this begins to open up a new way of thinking about our place in the universe.

Now, we might agree that there exists such enfoldment (after all, it is just standard physics under a Bohmian interpretation) but add that it is merely something passive and superficial. For example, surely the quantum field theoretical feature that each "particle" allegedly enfolds information about the whole universe applies to all "particles", and thus the fact that "particles" in my body do that is no more special than that the "particles" in, say, the table do that. And surely the effects of any such enfoldment must be negligibly small on the temporal and spatial scales in which we live our lives?

Now, the enfolded information that enters our sensory systems is, of course, more obviously relevant to us; but we might customarily think that even such enfoldment of information of, in principle, the whole universe through the senses is a passive and superficial relationship. Indeed, in traditional cognitive science, there has been a tendency to look at the human body as a machine that receives information through its sensory inputs, processes this information with the help of algorithms stored in the brain, and uses this information to behave in the physical world. In principle, all this has been thought to take place in a mechanical fashion, emphasizing that the environment, the information, and the brain/body have basically an external relationship with each other. Thus, one might think that a human being is basically a machine, and that the enfoldment relationship of this machine to the rest of the world is passive and superficial, not really affecting the inner nature of the machine.

Bohm, however, did not think that the enfoldment relationship between the part and the whole, and between the parts themselves, is merely passive and superficial. On the contrary, he emphasized that the enfoldment relationship is *active and essential to what each thing is*, implying that each thing is *internally related* to the whole, and therefore to everything else (Bohm 1990, p. 273).

What does "internally related" mean? *The Oxford Companion to Philosophy* tells us that the relation, R, between one item, x, and another item, y, is internal if x could not be the same item, or an item of the same kind, without standing in relation R to y (Bogen 1995, p. 756). Thus Bohm is implying that each thing – say, an electron, but also a human being – could not be what it is without standing in the enfoldment relationship to the whole universe. Instead, he proposed that the way the thing enfolds the whole is essential to what the thing is and to how it acts, moves, and behaves quite generally (Bohm 1987a, p. 41).

It is fairly easy to see what this means for human beings. Imagine, for example, losing your understanding that there exists a whole world of a certain kind of which you are a part. This would clearly make you into a very different person, and would probably profoundly change the way you "act, move, and behave" more generally. Of course, we take such understanding for granted and do not perhaps realize how fundamental it is for making us into what we are.

The more exotic suggestion is that even when it comes to an "elementary particle" of physics, such as an electron, the way it enfolds the whole is essential to what it is and how it acts, moves, and behaves.

It is well known that in certain situations at the quantum level of accuracy the electron behaves more like a wave, and in others more like a particle. Thus, according to some interpretations of quantum theory, what the electron is (i.e. whether it is a wave or a particle) is thought to depend on the nature of the environment that it interacts with (see, for example, Bohm's exposition of the "Copenhagen interpretation" of quantum theory (Bohm 1951)). The way the electron relates to the whole is thus thought to be essential to what it is, in a way that is very different from the situation in classical physics, where particles are what they are (i.e. particles), regardless of the kind of environment they happen to be located in. Also, quantum theory implies that tiny changes in the distant environment of the electron (e.g. the opening of a slit in a two-slit experiment) can have a profound effect upon its behavior, further underlining the dependence of the part upon the whole and the other parts. Furthermore, current experiments under so-called EPR conditions indicate that tiny changes in a state of a photon (e.g. those caused by a measurement of its polarization) can have an instantaneous effect on the behavior of another photon 50 km away (or at least there seems to be an influence between the regions that propagates much faster than the speed of light). A certain kind of wholeness is thus strongly implied by the behavior of matter in the light of quantum theory.

On the basis of such evidence coming from physics, Bohm proposes more generally that the whole is in a deep sense internally related to the parts. He adds that, since the whole enfolds all the parts, these latter are also internally related to each other, though in a weaker way than they are related to the whole (Bohm 1987a, p. 41).

Now, common experience tells us that there are also *external relationships* between things. In Bohm's terms, such external relationships are displayed in the *unfolded* or *explicate order*. The relation R between x and y is external if x stands in some relation R to y, but neither its identity nor its nature depends upon this being the case (Bogen 1995, p. 756). Think of a table and the various objects on it – the lamp, the book, the telephone, for instance. They are all related to each other. For example, they are outside one another, occupying different regions of space, at a certain distance from one another, each held to the table by the force of gravity, etc. However, removing one of the objects from the table will not change the identity or nature of the other objects. Another, larger-scale example of an explicate order is provided by the Solar System. And more generally, the physical world has a wide domain in which the explicate order prevails, all the way from the world of molecules to that of galaxies. At each level there are some entities that are relatively separate and extended.

Note that our conscious experience also has a domain in which an explicate order prevails. Think, for example, of your visual experience of the objects upon a table (as opposed to what exists "out there" independently of your experience). In the world of your visual consciousness, the explicate order typically dominates (cf. Honderich (2004)).

In the explicate order, each thing is thus seen as relatively separate and extended, and related only externally to other things. The explicate order dominates typical everyday experience, as well as classical (Newtonian) physics. It appears to stand by itself. However, Bohm proposes that it cannot be understood properly apart from its ground in the primary reality of the implicate order. This is an important point. The mechanistic world picture, based on classical physics, has assumed that the explicate order is all there is to the physical universe. In contrast, Bohm suggests that quantum and relativity theory show that the explicate order is merely a relatively autonomous order that has its ground in the more fundamental implicate order.

The next point Bohm makes is that the implicate order is not static but rather basically dynamic in nature, in a constant process of change and development. This is why he called its most general form the *holomovement.* The idea is that

> [a]ll things found in the unfolded, explicate order emerge from the holomovement in which they are enfolded as *potentialities*, and ultimately they fall back to it. They endure only for some time, and while they last, their existence is sustained in a constant process of unfoldment and re-enfoldment, which gives rise to their relatively stable and independent forms in the explicate order. (Bohm 1990, p. 273)

The above makes it clear that, as was already mentioned above, Bohm's ontology takes *movement* as fundamental, and here he connects with the tradition of "process philosophers" from Heraclitus to Whitehead (see, for example, Rescher (1996)). What does he mean by "movement"? Does he mean that there is something moving in the holomovement, some little particles or some little substantial fields that constitute *the* fundamental level of reality, and that the movement of these small-scale elements gives rise to the large-scale things (particles and fields) of everyday experience? No. He invites us to consider movement *per se* as fundamental, and things (e.g. particles and fields and whatever can be constructed from these) as derivative.

There are well-known problems with trying to take the notion of particles (whether point-like or extended) as fundamental:

> ... it is not possible in relativity to obtain a consistent definition of an extended rigid body, because this would imply signals faster than light. ... physicists were driven to the notion of a particle that is an extensionless point, but, as is well known, this effort has not led to generally satisfactory results, because of the infinite fields implied by point particles. Actually, relativity implies that neither the point particles nor the quasi-rigid body can be taken as primary concepts. Rather, these have to be expressed in terms of events and processes. (Bohm 1980, pp. 123–4)

(See also, for example, Bohm (1957, pp. 121–123).)

Such difficulties with the notion of particle have given rise to relativistic quantum field theory, and as we have already mentioned, today all matter is analyzed in terms of quantum fields, which are treated as the ground of all existence. However, Bohm and Hiley (1993, pp. 355–7) emphasize that these fields have to be understood not as some substantial entities in their own right that may or may not move, but rather as being *essentially* in movement. They say that, because of relativistic considerations, we can "never have the same field point twice", nor is there a unique form within the field that persists. Something with persistent identity would require a *unique relation* between the field point at one time and at other times. But if we consider the same field point over a period of time in different Lorentz frames, we do not see the same entity but different entities. This is why Bohm and Hiley emphasize that all properties that are attributed to the field have to be understood as *relationships in its movement.* The idea is that *the essential qualities of fields exist only in their movement.* It is not that there is a field with some essential qualities, which then may or may not move. Rather, it is movement that gives rise to the essential qualities of fields. Thus, movement is more fundamental, and the essential qualities (whether those of fields or particles) are derivative.

"Holomovement" then refers to the totality of such movement, which is assumed to be the most fundamental nature of existence known to us at present and which gives rise to the essential qualities of fields and particles. (As we have already mentioned, fields can give rise to particle-like manifestations via certain recurrent unfoldment and enfoldment.) If reality is more fundamentally *movement*, then the notion of a permanently existent entity with a given identity (e.g. a particle or a field with a unique form in it) is at best an approximation that works in a limited context (see Bohm and Hiley (1993, pp. 355–7)).

We have also seen that Bohm says that things *emerge* from the holomovement. But this is not a "something out of nothing" emergence or creation. Instead, Bohm assumes in an Aristotelian fashion that there exist *potentialities* in the holomovement. A potentiality for him is an "enfolded order" that "actualizes" when it unfolds to the explicate order. A thing that has actualized – say, an elementary particle such as an electron – then *endures*, but only for some limited period of time (for example, if the electron meets its antiparticle, the positron, they will both cease to endure as particles and instead transform into radiation). While a thing endures, it does not have a continuous existence as a particle-like entity. Instead, its existence is sustained in a constant process of unfoldment and re-enfoldment. Because such a process typically has *recurrence*, this gives rise to the relatively stable and independent form that we call the "particle" (Bohm 1990, p. 273).

If you like, the "particle" is a recurring *phase* of an underlying process of unfoldment and enfoldment. This, of course, is very similar to the idea of "incoming wave" Bohm had as early as in the late 1940s, in his first attempts to develop a more complete description of quantum reality. To get a better image of what this might mean, think of a spherical wave that closes in to a small region of space and then spreads out again, and think of another spherical wave (either a new wave, or else the original wave that is somehow reflected back) closing in to the same region in the next moment. At the moment when the wave has closed in and all the wave intensity is in the small region, the wave forms a "peak" that is very much like a particle (for example, it can transmit energy almost in a discontinuous, "all at once", fashion like a particle does, give rise to a localized spot on a photographic plate, like a particle would, etc.). And if there are waves closing in one after the other to the same region very rapidly, this approximates a situation in which there would be a particle just sitting there all the time. This is, roughly, the way one thinks of the fundamental mode of existence of a "particle" in the implicate order framework. And of course, as all the objects we find in everyday experience can be thought of as consisting of such "elementary particles" (e.g. electrons, protons, neutrons), then strictly speaking all objects can be seen as recurrent phases of an underlying movement of unfoldment and re-enfoldment.

Bohm's implicate order ontology contrasts with the ontology that has been prevalent in Western philosophy and science. This is the atomistic ontology, which assumes that everything consists of some fundamental elements (i.e. particles and/or fields) that are only externally related to each other. Atomistic ontology dominates much of contemporary science and philosophy (for example, it typically underlies the "mechanistic naturalism" Dennett refers to in the earlier quotation). But Bohm claims that physics strongly suggests that the atomistic ontology does not fit with the experimental facts of relativity and quantum theory. If he is correct, we need a new more fundamental ontology or theory of reality, and this is indeed what he tried to develop. He thought that the implicate order gives a valid and intuitively graspable account of the meaning of the properties of matter, as implied by quantum theory and relativity. He also thought that the implicate order framework can be extended to the domain of biological and psychological phenomena, making it into a proposal about the general architecture of existence as a whole, instead of just about physical existence.

Let us now move on to introduce Bohm's views about the nature of the mind and its relationship with matter, a topic that is the main focus of this book. He suggested that the implicate order applies even more directly and obviously to mind than it does to matter. In the mind, he tells us, there is

> a constant flow of evanescent thoughts, feelings, desires and impulses, which flow into and out of each other, and which in a certain sense, enfold each other (as, for example, we may say that one thought is implicit in another, noting that this word literally means "enfolded"). (Bohm 1990, p. 273)

"Constant flow" in this passage presumably refers to the "stream of consciousness". Bohm is then concerned with the order that prevails in this flow. If one assumes that only the explicate order prevails in the mind, then it would be natural to think that thoughts, feelings, desires, etc., are some sort of separate entities in mechanical interaction. Perhaps to some extent such analysis of the mind in terms of separate elements in mechanical interaction is adequate. However, Bohm suggests that such analysis has a very limited domain of applicability. For example, our thoughts flow into and out of each other. This suggests that a kind of enfoldment and unfoldment are the primary processes taking place in the stream of consciousness.

It thus seems that the implicate order prevails as the primary order of thoughts, feelings, desires, impulses, etc. However, in certain kinds of phenomenal consciousness, such as the visual experience of a static scene, it seems that the explicate order dominates. But there are aspects of even phenomenal consciousness where the implicate order seems to prevail. In particular, consider the structure of conscious experience over a period of time, or what we might call the temporal structure of consciousness or "time consciousness".

Consider, for example, what takes place when we are listening to music. We hear some notes now for the first time, but we also seem to actively perceive (rather than just passively remember) "past" notes, which the usual view of time says do not even exist. For the usual mechanistic view of time says that only things in the explicate order at the present instant of time exist, and other things (e.g. notes that were first heard some time ago) do not exist anymore. By introducing the notion of enfoldment, Bohm allows for a new kind of existence. Thus, he proposes that when I am listening to music, the past notes can exist as enfoldments, as active transformations of the original notes, and in this way they can be present and perceived in my conscious experience. The implicate order thus typically prevails in conscious experience. In contrast, when contemporary researchers discuss time consciousness, they tend to presuppose the view of time of classical physics (which exclusively emphasizes the explicate order). As a result, they find it very difficult to make any sense of the nature of our actual conscious experience, which seems to go beyond mechanical existence in the explicate order. As I will argue in more detail in later chapters, one of the potential benefits of the theory of the implicate order to contemporary philosophy of mind and consciousness studies is precisely that it provides these disciplines with one scientifically grounded possibility of becoming free from the unnecessary restrictions of classical physics. I suggest that those restrictions cause many of the troubles in philosophy of mind and consciousness studies. Classical physics is no longer the fundamental theory of matter, and thus there may be no valid scientific and philosophical reason to hold onto it as a tacit underlying framework, in the way many of those studying the mind today do.

Let us return to the more general discussion of the relationship between mind and matter. Because the implicate order also seems to prevail as the more fundamental order of the mind, Bohm was led to propose that the *general implicate process of ordering* is common to both mind and matter. This, he suggested, means that mind and matter are ultimately at least closely analogous and not nearly so different as they appear on superficial examination. Given the above analogousness, he thought it was reasonable to go further and indeed suggested in his 1980 book *Wholeness and the Implicate Order* that the implicate order may serve as a means of expressing consistently the actual relationship between mind and matter, without introducing something like Cartesian duality between them. However, he admitted that the implicate order, in the form proposed in his 1980 book, was at a relatively early stage of development. It should be seen as

> *a general framework of thought* within which we may reasonably hope to develop a more detailed content that would make progress toward removing the gulf between mind and matter. (Bohm 1990, pp. 273–4)

Bohm notes that even on the physical side, the theory about the implicate order

> lacks a well-defined set of general principles that would determine how the potentialities enfolded in the implicate order are actualized as relatively stable and independent forms in the explicate order. (Bohm 1990, p. 274)

He further notes that such a set of principles is also absent on the mental side. Even more importantly, he admits that the implicate order theory does not provide a clear idea of *just how* the mental and material sides are to be related. Therefore, if one wants to tackle the mind–matter problem more coherently, it is necessary to *extend* the framework of the implicate order.

Bohm first attempted such an extension by introducing a general theoretical notion he called "soma-significance" (Bohm 1985).[4] This notion tries to relate matter and meaning to each other, and if one assumes that meaning and the mental are overlapping concepts, this should also help to relate mind and matter to each other. The basic idea of soma-significance is that matter and meaning are not separate entities, but rather aspects of one overall reality. Thus, Bohm proposed that each particular significance or meaning is always based on some somatic order, arrangement, connection, and organization of distinguishable elements. For example, the printed marks on this page carry a meaning. When you perceive these marks, the result can be a meaning in your mind. But Bohm assumes, in accordance with contemporary cognitive neuroscience, that even that meaning in your mind is based on some somatic order in the more subtle levels of your brain.

He further suggested that reality, which is strictly speaking an undivided totality (indeed, the "holomovement"), can, for convenience, be thought of as being constituted out of relatively autonomous levels, which are organized into a hierarchy. There are manifest levels and there are more subtle levels, and each level has a somatic side and a

[4] Actually, Bohm does not explicitly say that the notion of soma-significance is an extension of the implicate order, but in my view it is natural to see it as such.

significant side. We are to think of the relation between the levels in terms of a process that has two directions: *soma-significant* and *signa-somatic.*

Soma-significant refers to the process in which the significance of a particular somatic order is carried over to higher levels, sometimes resulting in an apprehension of meaning in consciousness. Signa-somatic refers to the reverse process in which the significance that is apprehended acts "downwards" and organizes the less subtle levels. This, of course, is fairly similar to the notions of "enfoldment" and "unfoldment" in the implicate order framework. In the soma-significant direction of the process, information is gathered from the world and enfolded, as it were, and such enfolded information is carried over to higher levels where its meaning can be apprehended. When apprehended, the meaning of the information is unfolded, and as it unfolds, it can have an effect upon lower levels, which is the signa-somatic direction of the process.

To use Bohm's favourite example, think of the following situation: you are walking on a dark night and have heard that there is an assailant in the neighbourhood. Suddenly, you see a suspicious-looking shadow and interpret it as the assailant. Typically, this initiates a process that leads to somatic activity in the body in which the adrenalin flows, the heart beats, and one prepares to fight, freeze, or flee. In the example, the manifest level is that of the shadow. This is a somatic order that has a significance. This significance is then carried over, in a soma-significant process, to higher and more subtle levels in the brain, until finally its meaning is apprehended. This apprehension of meaning then typically gives rise to a signa-somatic process, via which the significance acts downwards to organize the somatic processes in less subtle levels.

The notion of soma-significance can clearly help us to think about how mind and matter are related. "Matter" corresponds to the more manifest levels, but it is assumed here that matter is not purely physical but that it always has a significant side, and in this sense a mental side, even though this may be fairly primitive. "Mind", in turn, corresponds to the more subtle levels, but it is likewise assumed that mind is not purely mental but that it always has a physical side, even though this may be very subtle. Because it is assumed that each level always has both somatic and significant aspects, it becomes possible to understand their relationship. Thus, it becomes possible to think that there is a "two-way traffic" between mind and body. The physical processes in the body influence the mind via soma-significant processes (which are enabled and indeed made necessary by the fact that soma is typically significant), while the mental processes in the mind affect the body via the signa-somatic processes (which are likewise enabled and made necessary by the fact that meanings typically "matter", or make a difference to the lower levels, by organizing them). Nowhere is there anything "purely physical" or "purely mental", and thus the traditional problem of how a "purely mental" mind can influence a "purely physical" body does not even arise.

Like the implicate order, the notion of soma-significance is a general scheme rather than a detailed theory. It is very useful for attempts to tackle the mind–matter problem, because meaning is clearly an important aspect of the mental, and soma-significance provides a general suggestion about how to think about the relationship between matter and meaning. It emphasizes that matter and meaning are not separate entities, but rather aspects of one reality, aspects that are present at each level of this reality. It also emphasizes that to understand the relationship between the mental and the physical, it is crucial to understand the relationship between matter and meaning. Matter, in general, has meaning, and thus it affects the mind. But the meanings apprehended in mind, in general, "matter", or make a somatic difference. Meanings are not just passive, abstract, separate entities, as our philosophical and scientific tradition often assumes, but rather they are seen as inseparable from the somatic aspects that underlie and ground them and which they in turn organize.

At the same time, something more specific needs to be said before a more detailed understanding of how mind and matter are related can be achieved. For example, the principle of soma-significance postulates that matter always has a significant side. But is it plausible that this is so, say, at the more fundamental levels of physics, and what could this mean more concretely?

We noted above that on both physical and mental sides, the theory about the implicate order lacks principles that would determine how the potentialities enfolded in the implicate order are actualized as relatively stable and independent forms in the explicate order. The notion of soma-significance as such does not provide us with such principles. Instead, it tries to capture how the implicate order (which can be seen as related to the more subtle levels) affects, via the signa-somatic process, the explicate order (which can be seen as corresponding to the more manifest levels), and how the explicate order (via the soma-significant process) can influence the implicate order. But it does not tell us how the manifest levels (explicate order) arise from the subtle levels (implicate order); it just presupposes that there are manifest and subtle levels.

We saw that a further difficulty with the theory of the implicate order was that although it goes some way toward helping us to understand the nature of and relationship between mind and matter, it does not provide a clear idea of *just how* mental and material sides are to be related. The notion of soma-significance says more about this. The manifest (explicate) levels can affect the subtle (implicate) levels because each level has both a somatic and significant side, and via the soma-significant and signa-somatic processes there can be a two-way traffic between manifest and subtle levels. If we assume that "matter" corresponds to the more manifest levels and "mind" corresponds to the more subtle levels, then we have a more precise way of thinking about how "matter" and "mind" are related and can mutually influence each other. The crucial point is that each level has both a material side and a significant (and in this sense a mental) side. But is it really plausible to claim that, say, matter at the quantum level has a significant side? More has to be said, in particular, about matter before the principle of soma-significance can really help to bridge the gulf between mind and matter.

Now, Bohm suggested that his ontological interpretation of quantum theory goes a long way toward extending the implicate order in the way required above (that is, it can provide a better view of how the potentialities are actualized, and how mind and matter are related). As we will see, it also provides a way of making the notion of soma-significance more specific at the quantum level of matter. Let us thus move on to consider this interpretation in a preliminary way (a more detailed description, involving some simple mathematics, will be provided in Chap. 4).

As we have seen, the ontological interpretation is based on an interpretation of quantum theory that Bohm originally proposed in 1952 and later developed especially in cooperation with his long-time colleague Basil Hiley (Bohm & Hiley 1993). Bohm presented in his 1952 papers both a "particle theory", that is, an ontological model of quantum particles (such as an electron), and, in an appendix, also a "field theory", that is, an ontological model of how the electromagnetic field of classical physics is "quantized" to give rise, at least momentarily, to "bullets of light" (i.e. quanta of energy, or photons), as quantum theory famously says it does.

Remember that Bohm felt that the ontological interpretation can do two things to make the implicate order more specific: firstly, to show

how the explicate order arises out of the implicate order, and secondly, to provide a more specific idea about how mind and matter are related. To see how the explicate order arises out of the implicate order, it is useful to consider the "field theory", that is, the ontological interpretation of the electromagnetic field. Roughly, one thinks of the electromagnetic field being in an implicate order (as we indeed mentioned above when saying that the movement of light waves in, for example, every region of the room enfolds information about the whole room). When one applies the ontological interpretation of quantum theory to this field, one then sees how the explicate order arises. The explicate order here is the famous "quantum", that is, a bullet of light, which in Bohm's theory has to be seen as a momentary, particle-like manifestation, rather than as a continuously existing particle. This, of course, is very much in the spirit of what we have said above about the implicate order. However, we will not discuss the ontological interpretation of the "quantized" electromagnetic field in more detail here but shall proceed to consider why Bohm thought that the ontological interpretation helps to understand more precisely how mind and matter are related. This is most easily seen by considering the "particle theory", that is, the ontological model of quantum particles (such as an electron).

As we have seen, according to Bohm's "particle theory" an individual quantum system (e.g. an electron) is always a combination of a particle and a new type of field described by the wave function ψ (so it is always both a particle and a wave, rather than either a particle or a wave, as one might say in the conventional interpretation of quantum theory).[5] If you like, the electron can be seen as an entity that has two aspects, a particle aspect and a wave aspect.

A useful analogy of Bohm's model of an electron is provided by a ship guided by a radar wave. In this analogy, the ship corresponds to the particle aspect of the electron, while the radar wave corresponds to the field aspect of the electron. Actually, the analogy is fairly good, because just as the radar wave influences the behavior of the ship, with the electron we can say that the field aspect influences the behavior of the

[5] Tarja Kallio-Tamminen (private communication) pointed out to me that for Niels Bohr, for example, "particle" and "wave" are classical concepts to be applied to classically observable phenomena, and thus he did not say that the electron is literally sometimes a wave and at other times a particle. However, it seems to me that other physicists fairly often think about the electron as a wave before it is observed and as a particle when it is observed (that is, the wave is commonly thought to "collapse" and in that way to give rise to a particle-like manifestation). In Bohm's ontological interpretation, however, there is no such collapse. The electron is always thought to be both a particle and a wave.

particle aspect. Furthermore, the form of the radar wave is determined by the shape of the environment (e.g. rocks in the bottom of the sea), and it is the form of the radar wave that is the key factor determining its influence upon the behavior of the ship. In an analogous way, the form of the quantum field is determined by the environment of the particle (e.g. the presence of various obstacles, slits, etc.) and, as we will see later in more detail, it is the form of the quantum field that is the key factor determining the influence of the field upon the behavior of the particle.

In Bohm's model, the way the field acts on the particle can be described by saying that the field gives rise to a new kind of potential energy, the "quantum potential", which in turn gives rise to a force upon the particle. The particle moves continuously along a trajectory, but now under the influence of the new kind of potential, the quantum potential, that the quantum field gives rise to. In conditions in which the effect of the quantum potential is negligible, quantum mechanics gives rise to Newtonian mechanics as a limiting case, as previously noted. In this way, one obtains an elegant means of resolving the notoriously difficult problem in quantum mechanics, namely the relation between the quantum level and the classical level. It can be argued that this model provides a clear and intelligible account of the movement of quantum particles, while avoiding the notorious paradoxes of quantum theory, such as wave–particle duality and the measurement problem (including the Schrödinger's cat paradox). Because the quantum potential typically gives rise to non-local correlations, it also makes explicit the striking feature of non-locality at the quantum level, a feature that has been demonstrated in a number of experiments since the 1980s (see, for example, Aspect et al. (1982)). On the whole, the ontological interpretation provides a clear and intelligible image of non-relativistic quantum phenomena.

Let us proceed to consider why Bohm thought that the ontological interpretation can be used to extend the implicate order in such a way that we obtain a better understanding of how mind and matter are related. We have seen that, according to this interpretation, an individual quantum system, such as an electron, is always a combination of a particle aspect and a field aspect, and the field influences the particle. The field gives rise to a potential, and from the potential one can calculate a force acting upon the particle. All of this may sound very close to the ideas of classical physics, as there are particles moving continuously and being pushed around by forces. Bohm and Hiley, however, emphasize that the ontological interpretation, although having some mechanistic features, also has new features that go radically beyond classical physics. In fact, this is not so surprising, because one obtains the ontological interpretation directly from the Schrödinger equation. Thus, one might expect that the new features of quantum theory will be carried over to the ontological interpretation. And because we now have an *ontological* interpretation of the mathematics of quantum theory, we might expect that the new, non-classical features of the mathematics stand out more vividly as aspects of the world (instead of being features of the formalism without our knowing their physical significance).

What are these new, non-Newtonian features of the ontological interpretation? When one looks at the mathematics describing the quantum potential, one sees something striking. The effect of the field on the particle *only* depends on the *form* of the field (while the effect of other fields in physics generally depends on the intensity of the field (= "size of the waves")). What does this mean? Bohm suggested that we have to look at the field as containing *information* that literally *informs* or puts form into the energy of the particle – we thus get a new notion of "active information". Active information as a general concept refers to a situation in which a form that carries very little energy enters into and directs a much larger energy. In a number of papers (including the one published in 1990 in the journal *Philosophical Psychology* (Bohm 1990)), Bohm used this idea as a basis for proposing the outline of a "new theory of the relation of mind and matter". The key idea is that there is a strong analogy between the way information in the quantum field acts on elementary particles and the way information in our subjective experience acts on the body. Just think of the above example of the shadow on the dark night. It is clearly the form of the shadow that is crucial for determining whether or not it will interpreted as meaning "assailant", with subsequent signa-somatic activities as a result. Analogously, it is the form (rather than the intensity) of the quantum field that determines its effect upon the particle. We see instances of active information also elsewhere: in the way the form of the DNA molecule is active in shaping the growth of a biological organism, in the way the form of the radio waves informs the energy of the radio receiver so that we hear a sound, in the way the form of the radar waves can guide the movement of the ship, in the way the information in a computer acts with various consequences, etc. Such information is clearly *objective* in the sense that it is primarily *information to the system*, rather than to us (that is, DNA is information to the cell, and the information contained in the wave function is information for the electron) (see Bohm

(1985, 1989, 1990), Bohm and Hiley (1993), Hiley (2003), Hiley and Pylkkänen (2001), and Pylkkänen (1985, 1992)).

Information contained in the quantum wave function has some holistic properties (for example, it "mediates" non-local connections), which makes it interesting to consider whether some of the well-known holistic properties of conscious experience could be connected with quantum active information. As we have already seen, Bohm hypothesized that mental processes are best understood in terms of a hierarchy of levels, each level having both a physical and a mental side, and where the more subtle levels organize the more manifest levels, while the more manifest levels provide content to the more subtle ones. The further proposal is to think that at each level, information is the bridge between the mental and the physical aspects.

Thus, when I consciously decide to move my hand and the hand moves, Bohm suggests that the information content in my conscious thought constitutes a subtle level of information that acts signa-somatically "downwards" in the hierarchy of levels, ultimately reaching the quantum level of information. Quantum-level information, in turn, acts on the elementary particles, atoms, and molecules (e.g. ions in synapses) or the electromagnetic field (for example, associated with the dendrites), and the effects of this can be amplified and result in a more classically describable physiological process, as a result of which the hand raises (see Hiley and Pylkkänen (2005)).[6] In a reverse process (e.g. in visual perception), the idea is that the incoming information is processed by the visual system first (mostly) in a classically describable way (where invariant features are abstracted etc.) up to a point where the information connects to the more subtle hierarchy of levels of information and ultimately the level of information that constitutes the content of conscious experience.

It is important to note here that the ontological interpretation, as proposed by Bohm in 1952 and Bohm and Hiley in 1993, does not take

[6] Notice that it is already commonly accepted that such amplification of the effects of individual quantum processes takes place in the early phases of vision, where a photon is absorbed by the 11-cis retinal molecule, causing it to change its shape. This effect is amplified and triggers a chain of events that first leads to a signal in the optic nerve, and eventually to a conscious experience of light (see Kandel, Schwartz, and Jessell (1991, pp. 404–5)) – remember, though, that it is not really understood how the physical process gives rise to a conscious experience. This makes it possible, at least in principle, that such amplification of quantum effects could take place elsewhere in the brain (assuming that the retina can be seen as a part of the brain). Note in particular that the retina works at 37 °C. It is thus not, in principle, necessary to have very low temperatures for such amplification of quantum effects in the brain.

into account any effect of the individual particle on its own quantum field (although they briefly sketch some ideas about how this might happen; see, for example, Bohm (1952a, pp. 171, 179); Bohm and Hiley (1993, pp. 345–6)). The idea that particles collectively affect the quantum field of a single particle is, however, contained in the standard notion that the shape of the quantum field of a particle is determined by the shape of the environment of the particle (which environment consists of many particles, and is part of the boundary conditions one puts into the Schrödinger equation before solving it, even in conventional quantum theory). The physicist Jack Sarfatti, in particular, has emphasized the need for an explanation of how the individual particle influences its own field and has proposed mechanisms for such "back-action", also emphasizing, in a very interesting way, its importance in understanding the mind–matter relationship and how consciousness arises (see, for example, Sarfatti (1997)).

Assuming that the notion of such an influence of the particle on its field can be coherently developed, we can then have two-way traffic between the mental and the physical levels without reducing one to the other. The role of Bohm's model of the quantum system then would be that it provides a kind of *prototype* that defines a more general class of systems in which a field of information is connected with a material body by a two-way relationship (a bit analogously to the way the Watt governor provides the prototype for the dynamical systems theory; see van Gelder (1997)).

Of course, what we have said above about active information connects with both the notion of the implicate order and that of soma-significance, when applied to the relationship between mind and matter. One question that was left open by these frameworks was just how mind and matter are connected, and in particular, how it is possible for mental processes to influence the more fundamental physical levels, if these latter are "purely physical". The proposal of active information at the quantum level makes it possible to address this question in a novel way.

We are now in a position to provide a brief summary of Bohm's way of thinking about mind and matter. In general terms, he saw mind and matter as two *aspects* of or *ways of looking* at an underlying reality, which is *movement*. This is a type of viewpoint that has roots in Aristotle and Spinoza and more recently in Russell, and is variously labelled "aspect monism" or "neutral monism" in philosophy. Of course, Bohm's emphasis on the fundamental status of movement connects him with the tradition of "process philosophy", from Heraclitus to

Whitehead. Aristotle's philosophy involves a dual aspect ontology and takes the notion of process as fairly fundamental (see Rescher (1986)). There is thus a particularly interesting similarity between Aristotle's and Bohm's views, also explicitly discussed by Bohm (see, for example, Bohm (1980, p. 12)).

Bohm further proposes that such reality can, for convenience, be analyzed in terms of levels that differ with respect to their subtlety and form a hierarchy. Each level then has both a physical and a mental aspect, and this makes "two-way traffic" between the levels possible. The levels are not separate entities in mechanical interaction. Instead, their relationship could be described as *mutual participation*. Participation has two sides, "to partake of" and "to take part in". A higher level partakes of a lower one, through its gathering of information about the lower one in a soma-significant process. But it also takes part in the lower level, by organizing it on the basis of what the information gathered means. Thus the levels in a sense enfold and unfold each other, and the implicate order prevails.

It has been mentioned many times that each level has both a physical side and a mental side. We have seen that Bohm suggested, radically, that even the quantum level can be thought to have, via active information, a mental side, a *primitive mind-like quality*, although he also thought it obvious that, say, an electron has no consciousness. I think that this is a very important new contribution that he made to mind–matter theory.

The idea that all parts of reality have a mental aspect is known in philosophy as panpsychism (see, for example, Nagel (1978)). To emphasize that the mental aspect associated with inanimate matter is very primitive, and that no full consciousness is attributed to all elements of reality, researchers have coined the term "panprotopsychism". Bohm's suggestion can be seen as an important contribution to panprotopsychism. Quantum theory is currently our most fundamental theory of matter, and Bohm suggests that quantum theory, when ontologically interpreted, reveals a proto-mental aspect of matter. This is the quantum field, described mathematically by the wave function, which is governed by the Schrödinger equation. This suggestion makes panprotopsychism a much more concrete scientific and philosophical proposal than it has hitherto been. Of course, one can always question Bohm's proposal, but he clearly gives some reasons to back up the idea that the wave function contains active information, and that active information in turn should be seen as a primitive mind-like quality (see Chap. 4). Of course, our mechanistic scientific and philosophical tradition goes strongly against attributing proto-mental qualities to the particles of physics. But when one looks at the history of science, one sees many instances where tradition has simply turned out to be mistaken. Thus, instead of just dismissing Bohm's suggestion as "obviously wrong", it might be more reasonable to fully consider the reasons he gives to substantiate his proposal.

What we usually call "mind" can then be seen as a fairly subtle level in the brain, with an internal relationship to the whole universe (through the implicate order). But like all levels, this level too has both a physical aspect and a mental aspect. Bohm assumed that for the "mind" the physical aspect is very subtle, for example more subtle than the quantum field (while in some respects similar to it). But the important point is that the mind is still assumed to have a physical aspect, and it can thus influence other such levels (e.g. the already-known neural levels) and be influenced by them. In this way, he claims to avoid dualism or idealism, without falling into reductive materialism. The whole point of double-aspect theories is, of course, to avoid these extremes.

Finally, we saw that Bohm further assumed that at each level, information is the *link* or *bridge* between the mental and the physical sides. In this way, he tried to answer the traditional objection against double-aspect theories or neutral monism, namely that it is left a mystery what is the nature of the reality of which mind and matter are thought to be aspects. (For a more detailed description, see, for example, Bohm (1990), Bohm and Hiley (1993, pp. 381–390), Pylkkänen (1992), Hiley and Pylkkänen (1997, 2001).)

1.3 An Overview of the Rest of the Book

We have now said enough about Bohm's views to be able to understand what the rest of the book does. His scheme is vast and ambitious, and it needs to be carefully thought about before one can judge whether it can actually provide us with the sort of better understanding of the nature of mind, matter, and their relationship that it tries to do. We need to go slowly and carefully, to explore the ideas, to criticize them, to digest them. Afterwards, we need to consider them in relation to other views, both to bring them into better focus and to see whether they provide any new understanding of the issues. Thus, the rest of the book divides into two parts. The first part (Chaps. 2–4) further explicates Bohm's views, while the second part (Chaps. 5 and 6) considers their relation to other viewpoints and the way they may tackle particular problems concerning the mind and its relation to matter.

I shall discuss Bohm's views in the order that best serves the purpose of understanding his views about mind and matter. The first challenge is to understand the implicate order framework. Chapter 2 examines the way this arises from quantum physics and relativity; the way it accounts for quantum features such as discontinuity of movement, wave–particle duality, and non-locality; the way it can be applied to cosmology; the way it enables one to think about laws of nature in general in a new way; and the way it might be extendable to biological phenomena.

In Chap. 3 we will explore Bohm's proposal that the implicate order is also the basic architecture of conscious experience. My discussion first focuses upon the justification for the suggestion that the implicate order prevails in the mind. This includes consideration of Pribram's holographic theory of memory in the brain; the phenomenal (e.g. spatio-temporal) structure of conscious experience of movement (such as the experience of listening to music or watching a motion picture, and the experience of movement more generally); consideration of the nature of the process of thought; and consideration of Piaget's ideas about the nature of the mind of very young infants. I will then move on to consider Bohm's suggestion about how matter and consciousness are related. The basic idea is that they are not interacting substances but rather correlated projections from a common ground. Their relationship would therefore in some ways be analogous to quantum non-locality, as interpreted by Bohm. I will suggest that this idea is related to Leibniz's famous ideas of "windowless monads" and "pre-established harmony", although there are differences as well. I will conclude the chapter by discussing the nature of time in the implicate order framework as well as the question of how strongly we should take Bohm's suggestions – are they meant as (almost) final truths, or as something more modest?

Chapter 4 then discusses how the ontological interpretation of quantum theory, with its notion of active information, might further enrich the picture, to give rise to a more comprehensive mind–matter theory. The subsequent two chapters will then compare and contrast Bohm's views with those of others in order to better clarify the various questions involved.

In Chap. 5 we fill focus on the paradox of time consciousness. This issue has recently been discussed by many researchers, and particularly extensively by the philosopher Barry Dainton (2000, 2001). I will argue that Bohm's views of the nature of the phenomenal structure of consciousness in terms of the implicate order provide a basis for a more adequate theory of time consciousness than those currently on offer, including Dainton's views.

In Chap. 6, I shall first further clarify Bohm's interesting but perhaps puzzling concept that reality is movement. I shall then consider whether minds have any genuine causal powers in the Bohmian universe. Finally, I shall conclude the book by briefly considering how one might address the so-called hard problem of consciousness in the Bohmian scheme.

2

The Architecture of Matter

2.1 Introduction

According to David Bohm, quantum and relativity physics point to a new notion of physical reality, where the key notion is that of implicate order. This notion, he further suggests, can be extended to the field of biological phenomena and consciousness, thus making it into a proposal about the general architecture of existence. In this chapter, our aim is to understand better the notion of implicate order as it applies in physics, and then in the next chapter to focus on how it might apply to the mind. But before focusing on physics, I think it is useful to briefly consider the more general background of the notion of implicate order. A good way to do this is to examine some quotations by Bohm in which he describes his project in very general terms. First, consider how he describes his work as a whole in the introduction of his 1980 anthology *Wholeness and the Implicate Order*:

> ... in my scientific and philosophical work, my main concern has been with understanding the nature of reality in general and of consciousness in particular as a coherent whole, which is never static or complete, but which is in an unending process of movement and unfoldment. (Bohm 1980, p. ix)

This quotation reminds us of important points, many of which were already made in the previous chapter. Firstly, Bohm's work has both a scientific and a philosophical dimension; this makes it richer but also harder to evaluate, given that science and philosophy are still often practiced in separation from each other. Secondly, he emphasizes understanding, over and above mere prediction and control, which often dominate in contemporary science, as it tends to focus on practical applications. Thirdly, there is the interest in our general concept of reality, and the proposal that wholeness, movement, and incompleteness are the more fundamental characteristics of existence. This contrasts with the prevalent atomistic concept of reality. It is particularly important to note Bohm's emphasis on the incompleteness of existence, for without it there would be no room at all for creativity in his concept of reality, as the unfoldment would be nothing but the realization of pre-existing enfolded potentialities. In contrast, incompleteness leaves room for the creation of new potentialities and new types of unfoldment, for example. Fourthly, there is a special emphasis on trying to understand the relationship between consciousness and reality – both the epistemic question about how consciousness can know external reality and the ontological question about what matter and consciousness are and how they are related.

Bohm recognizes that his project may sound too ambitious:

> These questions are, of course, enormous and could in any case probably never be resolved ultimately and completely. (Bohm 1980, p. x)

Why then did he bother to worry about them? He first suggested that such fundamental and deep questions are intrinsically interesting. This means that even though one could see no conceivable final answers or immediate pragmatic applications, it would still be fascinating to study the questions simply because this satisfies our intellectual curiosity and need to understand ourselves and our place in the broader scheme of things. But he also felt that there are other, more pragmatic and urgent, reasons why we ought to give attention to our general concept of reality:

> My suggestion is that a proper world view, appropriate for its time, is generally one of the basic factors that is essential for harmony in the individual and in society as a whole. (Bohm 1980, p. xi)

How could something so abstract and general as a world view be essential for harmony in the individual and in society? Bohm suggested that the way we think about the totality influences the way our minds tend to operate, which in turn influences our actions. Thus, an incorrect or confused view of the totality might facilitate a confused operation of the mind and confused actions. In particular, he felt that our prevalent modern scientific view of totality, deriving from classical physics, gives too much emphasis to the idea that reality consists of independent parts. This overly divisive view tends to give rise to an overly divisive operation of the mind. In the social sphere such overly divisive thought easily leads to egoism, that is, putting the interests of the parts (especially the needs of one's own ego) first and those of others and the whole second.

Bohm argued that quantum physics and relativity suggest another view of the totality which underlines the undivided wholeness of the universe. He felt that such a holistic view of totality could help to bring about a more holistic operation of the mind, leading to more orderly action within the whole. There is thus a strong social component in his view of metaphysics, evident also in his interest in group dialogue as a kind of "sociotherapy" (Bohm 1985; 1995).

Of course, it is not obvious that Bohm is correct in his social analysis. For example, perhaps a certain amount of egoism at the "microlevel" is necessary for harmony at the "macrolevel", and too much emphasis upon the whole easily becomes totalitarian and stifles individual creativity. Yet we can witness many instances in past and current politics where it would clearly have been wiser to put the interests of the whole first. To give a recent example, think of the difficulties of getting different countries to sign the Kyoto global warming agreement. I think the question about the possible social implications of our general world view both an interesting and important one, but it will not be the main focus of this book.

Let us thus continue to discuss Bohm's proposed holistic concept of reality. It is useful to consider another quotation in which he describes his project in general terms, this time in the introduction to the seventh and final chapter of *Wholeness and the Implicate Order*:

> Throughout this book the central underlying theme has been the unbroken wholeness of the totality of existence as an undivided flowing movement without borders. It seems clear... that the implicate order is particularly suitable for the understanding of such unbroken wholeness in flowing movement, for in the implicate order the totality of existence is enfolded within each region of space (and time). So, whatever part, element, or aspect we may abstract in thought, this still enfolds the whole and is therefore intrinsically related to the totality from which it has been abstracted. Thus, wholeness permeates all that is being discussed, from the very outset. (Bohm 1980, p. 172)

The phrase "unbroken wholeness in flowing movement" summarizes Bohm's concept of reality fairly neatly. It emphasizes that reality is an unbroken whole and that reality is movement. To say that reality is an unbroken whole is to deny atomism, the view that reality fundamentally consists of some basic elements, such as particles or fields. To say that reality is movement is to deny the view that reality consists of some static things that may or may not move. Instead, movement is assumed to be fundamental, and things are assumed to be derivative, for example relatively invariant patterns in the movement, analogous to vortices in a stream.

Bohm's view is exotic. For in everyday sensory experience we typically meet a world that is made up of its parts, which often just sit where they are. Think of the objects in your table. How does Bohm reconcile the static and broken wholeness of our everyday experience with his proposal about reality as "unbroken wholeness in flowing movement"? As we saw in the previous chapter, he agrees that the static and broken wholeness we meet in everyday experience is a real aspect of reality, a part of what he calls the "explicate order". However his key proposal is that the explicate order, although autonomous, is only relatively so. More fundamentally, the implicate order prevails in reality. This means that, strictly speaking, the totality of existence is enfolded in each region of space and time. We can abstract a part, an element, or an aspect from reality, but because the implicate order is the fundamental order of reality, such a part, an element, or an aspect will, strictly speaking, inevitably enfold the whole and thus remain intrinsically related to the totality. Wholeness thus permeates, from the very beginning, all that we can abstract and discuss. This is exotic, but not unheard of. For example, the poet and artist William Blake famously captured something similar when he wrote:

> To see a World in a Grain of Sand
> And Heaven in a Wild Flower
> Hold Infinity in the palm of your hand
> And Eternity in an hour
> (from *Auguries of Innocence*, c. 1800)

Bohm's basic architectural proposal thus is that the totality of existence (which is "unbroken wholeness in flowing movement") is enfolded within each region of space and time. It is obvious that such a change in our general conception of reality can – at least potentially – have strong implications for our understanding of a wide range of phenomena. In this book we are concerned with how the change of our general conception of reality in a Bohmian fashion might influence our understanding of the nature of mind and its relationship to matter. As we saw in Chap. 1, Bohm himself was well aware of such possibilities, and indeed proposed the outline of a new theory of the relationship between mind and matter, and also characterized the temporal and spatial structure of conscious experience in a novel way.

He provided a non-technical presentation of the main features of the implicate order in the final chapter of *Wholeness and the Implicate Order*. This presentation is very useful for our purposes, and we shall be partly following and discussing it in this chapter and the chapter that follows. Bohm's presentation begins by bringing out how the implicate order arises from quantum and relativity physics, and how it can be used to discuss the general architecture of matter, first at the atomic level but also in cosmology. The basic idea is that certain new features of physical processes which seem impossible to comprehend in terms of the traditional mechanistic order can be coherently discussed in the implicate order framework. These features include discontinuity of motion (the famous "quantum jumps"), wave–particle duality, and non-locality. Bohm then tried to show how the implicate order can be extended to the field of biological phenomena and consciousness, thus making it into a proposal about the nature of reality in general, rather than just the physical aspect of reality. We can thus see the emergence of a new proposal about the architecture of being. This involves a new way of thinking about some key categories, such as causality, space, time, and movement. These categories will be crucial when we try to discuss the nature of mind and its relation to matter in the context of Bohmian ontology.

2.2 The Role of the Notion of Order in Physics

We saw in Chap. 1 that the implicate order involves thinking about a particle, such as an electron, as a set of ingoing and outgoing waves. The particle-like aspects of the electron can then be seen as recurrent phases of such an underlying process of unfoldment and enfoldment. Bohm tells us that the germ of this idea of the electron as a set of ingoing and outgoing quantum mechanical waves occurred to him as early as about 1950, when he was in Princeton and had began to look for a more complete model of reality at the quantum level (see Bohm (1987, pp. 34–35); see also Pylkkänen (1996b)). Thus he was anticipating some aspects of the implicate order already before he published, in 1952, the ontological interpretation of quantum theory, which relies more heavily on the concept of a particle. However, the focus on "order" becomes more important in the 1960s, partly as a result of a long and fruitful correspondence with the American artist Charles Biederman, who was deeply concerned with the role of order, structure, and process in nature (see Bohm and Biederman (1999); Pylkkänen (1999b)). Important articles that bring out Bohm's early ideas about order include his inaugural lecture "Problems in the Basic Concepts of Physics", delivered at Birkbeck College, University of London in February 1963, and published with two appendices in 1965 (Bohm 1965b), and the paper "Space, Time, and the Quantum Theory Understood in Terms of Discrete Structural Process", which was also published in 1965 (Bohm 1965c).

In his 1965c paper, Bohm's starting point is to note that in the past half century there had been a series of revolutionary developments in physics, which include relativity theory, quantum theory, and the theory of elementary particles. There had notoriously been fundamental contradictions and ambiguities in the efforts to bring together these theories, which suggested that basically new ideas were needed to obtain such unification and a better understanding of the situation.

His proposed new line of development began by suggesting that we should give up the notion of continuous space-time as fundamental and replace it by the notion of space-time as a *discrete structural process.* The rough idea here is that "structural process" refers to a set of "space-like" elements. These are discrete structures which undergo discrete or discontinuous changes as they move and unfold in a process of development. Continuous space-time can then be seen as an abstraction from the underlying discrete structural process (Bohm 1965c). Note that one here uses the word "process" to refer not to a continuous change but rather to a discrete, step-by-step change. Indeed the word "process" is based on the verb "to proceed", which means "to step forward". It originally thus refers to a particular kind of movement, which goes step by step, with one step following another (Bohm 1976, pp. 40–41). This is the sense in which Bohm uses the word process here, which perhaps makes the term "discrete structural process" easier to grasp.

To develop the notion of "discrete structural process", Bohm thought is was necessary to begin by going axiomatically to the notion of *order.* This contrasts with those approaches in physics which begin by postulating a certain "fundamental length". The reason he did not want to begin by taking length or measure as fundamental is that he thought that *every theory of measure or length tacitly presupposes a notion of order.* His suggestion here was that if one does not know what is to be meant by the order of points, the measure of the distance between them can have no significance. Thus order is a more fundamental concept than that of length or measure, and consequently Bohm thought that in search of a new fundamental theory in physics we ought to begin by focusing on the notion of order. His strategy was therefore first to explicate the notion of order, and then from this basis to try to derive length or measure as a higher-level concept.

Of course, the above may sound very abstract, and perhaps some concrete examples would help here. Think about trying to build a house. Here it is useful to have some elements of the same size, such as bricks. The order of such elements then defines the structure of the house. Similarly, think of a musical composition. Here one typically has some elements of well-defined size and properties (notes of different pitch and duration, for example), and the order of such elements defines a composition. Analogously, to understand the structure of the universe it might be useful to understand the nature of the basic elements and the order that prevails amongst them. In classical physics the basic elements were assumed to be atoms and fields, and the universe was thought to be a huge clockwork, a structure that is defined by a certain order of the particles and fields, and the laws that determine how the structure changes over time.

Bohm turns the familiar mechanistic scheme upside down. Instead of saying that the universe is made of some basic elements (particles and/or fields), he postulates that the universe is movement, "all is flux". Order is then something that prevails in this movement, and one of the consequences of this is the set of "space-like elements" that undergo discrete or discontinuous changes. Continuous space-time is then something we abstract from this underlying discrete process. Thus we do not construct the universe by starting with some basic elements in space-time. Instead, we start with flux and the idea that order prevails in the flux. This gives rise to elements which can be ordered into the familiar structures. The structure of the universe thus depends on order even in the Bohmian scheme; but according to Bohm, quantum theory and relativity inspire us to start thinking about "order in movement" as fundamental, and the order of things (the explicate order) as derivative.

Having arrived at the idea that order is a fundamental notion in the current situation in physics, Bohm also began to consider the role of the notion of order in previous revolutionary changes in physics. The Ancient Greeks, most notably Aristotle, had a certain kind of physics, which was replaced by the classical physics of Galileo and Newton at the beginning of the modern era. Further developments within classical physics involved Maxwell's electromagnetic theory, the molecular theory of heat, and the kinetic theory of gases. These gave rise to relativity theory and quantum theory, which in turn became the basis of theories such as quantum electrodynamics, quantum field theory, and quantum chromodynamics, all this leading more recently to string theories etc. But, Bohm asked, what is it that changes when a new theory arises?

He suggested that there are essentially two things that happen. On the one hand, a *perception of a new order* takes place. On the other hand, people try to develop *new ways of using language* that are appropriate to the *communication* of such a new order. So the idea is that changes in physics fundamentally involve *perception of a new order* and *communication of a new order.*

The key point in our own age is that in the quantum and relativistic context, the old mechanistic orders of classical physics have ceased to be relevant. In other words, they can no longer coherently be adapted to fit the experimental facts of quantum and relativistic phenomena. However, Bohm thought this crucially important point has not been given sufficient attention by the majority of physicists. One example of this is the still prevalent use of Cartesian coordinates in physics, including quantum physics. These coordinates are particularly suitable for describing the mechanistic order of classical physics. But, as already mentioned, the mechanistic order does not fit the facts of quantum and relativistic phenomena. This strongly suggests that we should question the use of Cartesian coordinates to describe quantum and relativistic phenomena, and underlines the need to develop some new, more fitting ways to describe them. But to succeed in this, Bohm thought, we first need to perceive a new order and then develop new ways of using language, including new mathematical forms, to communicate such a new order. This is what he set out to do, and the result was the proposal that the implicate order is the new order to which relativity and quantum theory are pointing.

Note that the Cartesian order is good at describing the domain of classical physics. What we need, however, is a deeper framework in which the Cartesian order can be embedded. Consider, for example, the famous "quantum jump". This means that the electron momentarily disappears from the Cartesian coordinate system, as it were, only to appear some distance away from the point where it disappeared. But what happened in between? This can naturally be described in another framework, such as the implicate order framework, with its notions of enfoldment and unfoldment.

We will now move on to see in more detail how Bohm was led to the notion of the implicate order. The first question to consider is how relativity and quantum theory challenge the traditional mechanistic order.

2.3 Relativity, Quantum Theory, and the Mechanistic Order

We have already noted that in physics there has been a succession of ideas of order. For example, the Ancient Greeks had a certain idea of order in nature that emphasized the order of degrees of perfection of celestial matter, which corresponded to the order of distance from the centre of the Earth (Bohm 1980, p. 112). With the development of classical physics, the relevance of the Greek notion of order was questioned, and a new *mechanistic* notion of order was perceived and developed. This mechanistic order simply assumes that the world is

- constituted of entities (particles and/or fields) which are *outside* of each other, in the sense that
- they *exist independently* in different regions of space (and time) and
- interact either via direct push and pull or else through forces that do not bring about any changes in their essential natures. (Bohm 1980, p. 173)

The mechanistic order clearly emphasizes the external relationships between things, indeed so much so as to imply that there are only external relationships (see Chap. 1 for a brief discussion of the nature of external and internal relations). The above, mechanistic description of the world may sound obviously correct, and is often presupposed even in contemporary metaphysics and philosophy of mind (see, for example, Dennett (1999)). Yet developments in contemporary physics challenge the idea that this description is fundamental, and it is precisely this which makes contemporary physics philosophically so important.

Bohm often used the metaphors of machine and living organism to illustrate the difference between a mechanical order and a non-mechanical or more holistic order:

> The machine gives a typical illustration of such a system of [mechanistic] order. Each part is formed (e.g., by stamping or casting) independently of the others, and interacts with the other parts only through some kind of external contact. By contrast, in a living organism, for example, each part grows in the context of the whole, so that it does not exist independently, nor can it be said that it merely "interacts" with the others, without itself being essentially affected in this relationship. (Bohm 1980, p. 173)

Thus, parts of a living organism are internally related to each other, while parts of a machine are only externally related to each other. When it comes to the physical world of inanimate matter, what is the order that prevails there? Classical physics assumed that this is the mechanistic order, and the mechanistic order indeed works approximately well for the domain of classical physics. Bohm emphasizes, however, that both relativity and quantum theory challenge the relevance of the mechanistic order. Thus, if he is correct, a scientific ontology has to give up the mechanistic order as fundamental.

The essence of the mechanistic order is, of course, the idea that there exists a fundamental level of reality which consists of elementary particles. As we saw in the introduction, relativity implies, however, that we cannot have a coherent concept of an independently existent particle, neither one in which the particle would be an extended body, nor one in which it would be a dimensionless point (Bohm 1980, p. 173). Bohm points out that in order to meet these difficulties, Einstein attempted to develop a *unified field theory*, in which the particle concept is not taken as primary but reality is instead regarded as constituted of fields that obey relativistic laws and give rise to particle-like manifestations as regions of intense field. Although Einstein was not able to complete his program of unified field theory, Bohm thinks it important because it did suggest a concept of reality as an unbroken and undivided totality from which the particle concept may be achieved as an abstraction.

Bohm emphasizes that the quantum theory presents a much more serious challenge to the mechanistic order than does the theory of relativity. He lists three key features of the quantum theory that challenge mechanism. First of all, movement is thought to be in general *discontinuous*. This means that a physical quantity known as action is constituted of indivisible quanta. A famous implication of this is the so-called "quantum jump". An electron, for example, is thought to be able to go from one state to another without passing through any states in between.

Second, there is the phenomenon of *wave–particle duality*: individual quantum systems such as electrons, which classically were assumed to be just particles, can exhibit different properties (e.g. particle-like, wave-like, or something in between), depending on the environmental context within which they exist and are subject to observation.

Third, there is the feature of *non-locality*. Two systems, such as electrons, that initially combine to form a molecule and then separate show a peculiar non-local relationship, which Bohm suggests can best be described as a "non-causal connection" of elements that are far apart.

How do these features challenge the mechanistic order? Bohm answers:

> ... if all actions are in the form of discrete quanta, the interactions between different entities (e.g., electrons) constitute a single structure of indivisible links, so that the entire universe has to be thought of as an unbroken whole. In this whole, each element that we abstract in thought shows basic properties (wave or particle etc.) that depend on its overall environment, in a way that is much more reminiscent of how the organs constituting living beings are related, than it is of how parts of a machine interact. Further, the non-local, non-causal nature of the relationship of elements distant from each other evidently violates the requirements of separateness and independence of fundamental constituents that is basic to any mechanistic approach. (Bohm 1980, pp. 175–6)

Quantum theory thus seems to require fundamentally a holistic rather than a mechanistic concept of reality. The experimental results clearly suggest that there is an internal relationship between an electron and its environment, or between two electrons in a non-local relationship. Our familiar reality at the macroscopic level obeys the principles of the mechanistic order, with external relationships, but only approximately. In a more accurate description, the new quantum physical principles, and thus a new holistic order, seem to prevail. But what kind of holistic order? This is what Bohm set out to find out, and this search led him to the notion of the implicate order.

2.4 From the Mechanistic Order to the Implicate order

Bohm next points out that although both quantum theory and relativity challenge the mechanistic order, their basic concepts directly contradict each other. Relativity requires continuity, strict causality (or determinism), and locality, while quantum theory in its conventional interpretation requires non-continuity, non-causality, and non-locality. He comments:

> It is therefore not surprising that these two theories have never been unified in a consistent way. Rather, it seems most likely that such unification is not actually possible. What is very probably needed instead is a qualitatively new theory, from which both relativity and quantum theory are to derived as abstractions, approximations and limiting cases. (Bohm 1980, p. 176)

But how does one go about developing such a theory? Here's how he suggests we proceed:

> The basic notions of this new theory evidently cannot be found by beginning with those features in which relativity and quantum theory stand in direct contradiction. The best place to begin is with what they have basically in common. This is undivided wholeness. Though each comes to such wholeness in a different way, it is clear that it is this to which they are both fundamentally pointing. (Bohm 1980, p. 176)

To develop the new theory thus requires that one start with some concept of undivided wholeness, instead of the mechanistic order that has been customary in physics. However, Bohm is well aware that it is not easy to give up the mechanistic order as fundamental in favor of some other notion of order that describes undivided wholeness. He notes that the mechanistic order has, for many centuries, been basic to all thinking in physics. This reflects itself also in the central role that the Cartesian coordinates – which are particularly well suited to expressing the mechanistic order – still play in physics. Physics, as we have seen, has changed radically in many ways, not least in challenging the mechanistic order. Yet, Bohm points out, the Cartesian grid (with minor modifications, such as the use of curvilinear coordinates) has not changed. He points to reasons making any such change difficult:

> Evidently, it is not easy to change this, because our notions of order are pervasive, for not only do they involve our thinking but also our senses, our feelings, our intuitions, our physical movement, our relationships with other people and with society as a whole and, indeed, every phase of our lives. It is thus difficult to "step back" from our old notions of order sufficiently to be able seriously to consider new notions of order. (Bohm 1980, p. 176)

So this is how Bohm sees one of the main problems facing contemporary physics, and I think this is also a problem that those contemporary philosophers who are trying to develop a physicalist ontology which tries to do justice to quantum theory and relativity are facing. On the one hand, the experimental evidence coming from relativity and especially quantum physics strongly suggests that the mechanistic order is inadequate as a fundamental characterization of the architecture of the physical world. On the other hand, the mechanistic order is the one that physicists adopted hundreds of years ago and that strongly pervades their thinking even today. As a result, physicists typically try to deal with the undivided wholeness of quantum and relativistic phenomena in terms of the mechanistic order. Mathematically, this shows itself in the prevalence of the Cartesian grid. But it also shows itself, for example, in the way physicists still often talk in terms of "elementary particles", thus implying that the essence of physics is to seek the ultimate constituents of matter, in the spirit of the mechanistic order. But how, for example, does one reconcile the focus on "elementary particles" with the relativistic difficulties concerning the concept of particle in the first place, or with the discontinuity of motion, or with the wave–particle duality of quantum interference experiments, or the non-local correlations between such "particles"?

If the aim of physics is to provide us with a better understanding of the fundamental nature of the physical world (as opposed to merely providing tools for technological applications), then it is clearly not adequate to hold on to the mechanistic order, if experimental evidence strongly exhibits a fundamentally different kind of order, such as the order of "undivided wholeness". Yet if Bohm is correct in saying that the mechanistic order fundamentally pervades our minds, it is not a trivial task to give it up. For if the mechanistic order is truly pervasive, then any attempt to do any thinking at all (including the very attempt to get rid of it) might in some subtle way be controlled by the mechanistic order from the very beginning! Thus even if a physicist facing the non-mechanistic experimental evidence sets out to develop new notions of order, if she or he is not aware of the pervasive nature of the mechanistic order, this order is likely sooner or later to "take over" the new theory, with confusion as a result. Perhaps something like this has happened in much of contemporary physics?

How does Bohm then suggest that we proceed? If our current situation is that the mechanistic order fundamentally pervades our minds, then we have no choice but in some sense to start from it. Consequently, his strategy is to consider examples, models, and analogies that are in a sense mechanistic (and thus easily understandable to our minds pervaded by the mechanistic order) and yet are able to *illustrate* something non-mechanistic, some new holistic principles. He chooses examples that directly involve sense perception, as well as models and analogies that try to illustrate his new notion of order in an imaginative and intuitive way. But it is very important to realize at the outset that such analogies should be seen merely as *tools* that in a sense have to be *thrown away* once they have done their job. One thinks here of Wittgenstein, who at the end of the *Tractatus* writes:

> My propositions are elucidatory in this way: he who understands me finally recognizes them as senseless, when he has climbed out through them, on them, over them. (He must so to speak throw away the ladder, after he has climbed up on it.) He must surmount these propositions; then he sees the world rightly. (Wittgenstein 1974, paragraph 6.54)

Analogously, to understand what Bohm is trying to convey with his mechanical analogies, it is important to "kick away the ladder" from time to time.

Before discussing some analogies that illustrate the order of undivided wholeness, let us first follow Bohm in considering an instrument that has provided support for the mechanistic order, namely the photographic lens. His idea is that

> the photographic lens is an instrument that has given us a very direct kind of sense perception of the meaning of the mechanistic order, for by bringing about an approximate correspondence between points on the object and points on the photographic image, it very strongly calls attention to the separate elements into which the object can be analyzed. By making possible the point-to-point imaging and recording of things that are too small to be seen with the naked eye, too big, too fast, too slow, etc., it leads us to believe that eventually everything can be *perceived* in this way. From this grows the idea that there is nothing that cannot also be *conceived* as constituted of such localized elements. Thus, the mechanistic approach was greatly encouraged by the development of the photographic lens. (Bohm 1980, p. 177)

By drawing attention to the role of the photographic lens, he is referring to one possible reason why the mechanistic order has become so prominent. It is not just that the lens emphasizes analyzability. More importantly, because the lens has been so successful in many areas, we make the generalization that it will work in *all* areas. Thus we might make the assumption that for something to *exist at all* as a physical entity, it has to be *perceivable* with the help of the photographic lens, and thus be *analyzable* into its elements. Bohm suggests that this assumption about perceivability further gives rise to assumptions about *conceivability*, which are even stronger assumptions. For if we assume that it is not even conceivable that there exists something physical that is not constituted of localized elements, we have indeed adopted the mechanistic approach in a powerful way.

The point here is that our scientific instruments can affect our assumptions about perceivability; and these in turn can affect our assumptions about what is conceivable. In this way, a scientific instrument such as the photographic lens can have a profound effect upon the way we *think* about the world at the most fundamental and general level, that is upon our general world-view, metaphysics, or ontology.

Relativity and quantum theory suggest that the mechanistic order is inadequate; and yet the scientific tradition, powerfully backed up by instruments such as the lens, encourages us to think that there is nothing that cannot be perceived, or even conceived, in the mechanistic way. Bohm's strategy here is to consider another instrument, which draws our attention to something different, namely to undivided wholeness, analogous to the way the photographic lens drew our attention to the primacy of parts. This other instrument is the *hologram*.

In a hologram, each part of the photographic plate contains information about the whole object so that there is no point-to-point correspondence of the object and the recorded image. A hologram makes a photographic record of the interference pattern of light waves that have come off an object. The form and structure of the entire object can thus be said to be *enfolded* within each region of the photographic record. When one shines light on any region, this form and structure are then *unfolded*, to give a recognizable image of the whole object once again (Bohm 1980, p. 177). Bohm emphasizes that a *new notion of order* is involved here, which can be called the *implicate order* (the verb "to implicate" means "to fold inward" or "to enfold"). His idea is that in terms of the implicate order one may say that "everything is enfolded into everything". This contrasts radically with the *explicate order* now dominant in physics in which things are *unfolded* in the sense that each thing lies only in its own particular region of space (and time) and outside regions belonging to other things (Bohm 1980, p. 177).

Just as the lens affected our ideas about perceivability and conceivability, so can the hologram. The lens, according to Bohm, encouraged the idea that it is not even conceivable that there are physical objects not made up of their parts or basic elements. The hologram, however, draws attention not only to the parts but also to the whole. Thinking about the hologram can thus challenge the deeply held assumptions about perceivability and conceivability that the lens has strengthened. In particular, a new *part–whole* relation becomes conceivable, the idea that the whole is in some sense contained in each part, and that each part, in turn, is enfolded throughout the whole. This suggests a new idea of the architecture of reality. The hologram thus considerably helps us to conceive a new order, the implicate order, just as the lens helped us to better perceive and conceive the nature of the explicate order.

One might note here that in the history of philosophy there has been a tendency to appeal to conceivability as an important criterion when dealing with philosophical problems. Consider David Chalmers:

> There is a long tradition in philosophy of using *a priori* methods to draw conclusions about what is possible and what is necessary, and often in turn to draw conclusions about matters of substantive metaphysics. Arguments like this typically have three steps: first an epistemic claim (about what can be known or conceived), from there to a modal claim (about what is possible or necessary), and from there to a metaphysical claim (about the nature of things in the world). (Chalmers 2002, p. 145)

If we consider Bohm's remark about the role of the lens as support for the mechanistic view of nature in the light of Chalmers' comment above, we could say that the lens first gave rise to epistemic assumptions (we can only perceive and conceive things that are constituted of localized elements); then to modal assumptions (it is not possible for there to be physical entities that are not constituted of localized elements); and finally to metaphysical assumptions (the world is made up of localized elements; that is, the explicate order is the fundamental order of the world). In contrast, the hologram suggests that we can easily conceive how the whole can be enfolded in each part, and likewise how the part can be enfolded all over the whole. This can encourage us to make the modal assumption that there *can be* physical phenomena that are not constituted of localized elements but that instead the whole can be enfolded in each part and the part enfolded in the whole. And finally, given the experimental evidence coming from quantum and relativity physics, we could with some justification make the metaphysical assumption that some such holistic order *actually prevails* as the most fundamental order of the world, as it is known today.[1]

The hologram is extremely valuable for Bohm's project because it helps to illustrate the idea of the implicate order in a sensibly perceptible way. But he goes on to remind us that it is only an instrument whose function is to make a static record (or "snapshot") of this order.

[1] For recent suggestions that holographic principles prevail in fundamental physical processes, see, for example, Bekenstein (2003) and Maldacena (2005). Note also how Bohm's discussion suggests that conceivability need not always be an a priori issue. For example, in so far as the lens has influenced what we find conceivable, sensory experience has influenced conceivability, and thus conceivability becomes at least partly an a posteriori issue; that is, it depends partly upon experience.

The order that is being recorded is the *order in movement*, namely in the complex movement of electromagnetic fields, in the form of light waves.

> Such movement of light waves is present everywhere and in principle enfolds the entire universe of space (and time) in each region (as can be demonstrated in any such region by placing one's eye or a telescope there, which will "unfold" this content. (Bohm 1980, p. 177).

Thus we do not really need a hologram to get an example of the implicate order. As mentioned in Chap. 1, the implicate order is there right in front of you, wherever you go, for it is the order that prevails in the movement of electromagnetic fields. Information about the whole universe can be enfolded in a small region of space, and this information can be unfolded by a suitable instrument. To consider a more down-to-earth example, when one is in a room, the light waves in each region typically enfold information about the objects in the whole room. Thus information about the "explicate order" of, say, the furniture in the room is stored in an implicate form in the movement of light waves in each region of the room.

To underline the universality of the implicate order, Bohm points out that such enfoldment and unfoldment takes place not only in the movement of the electromagnetic field, but also in that of other fields, electronic, protonic, sound waves, etc. Thus whenever we listen to something, we are making use of the implicate order. The information carried by sound waves is typically enfolded in each region (including, of course, the region inside the ear), analogously to the way light enfolds information. But quantum field theory, as seen by Bohm, implies a yet more radical example of the implicate order. We are perhaps used to thinking about atoms and electrons as little billiard balls. Quantum field theory, as interpreted by Bohm, suggests that we instead describe an electron primarily as a field activity, and in that activity the implicate order prevails, just as it prevails in light and sound waves. This suggests that the implicate order is very fundamental indeed. For the idea is that even the "particles" that constitute, say, the table in front of me (or my brain, for that matter!) are, in a more accurate description, stable patterns in fields which enfold information about the whole universe!

Science has already discovered a great number of fields in which the implicate order prevails. However, Bohm thinks it is likely that many additional ones, as yet unknown, may be discovered in the future. The task of the laws of physics is then to describe the behavior of such fields. Classical physics (e.g. Maxwell's equations for the electromagnetic field) provides an approximate description of this behavior, but more accurately the laws of quantum theory are needed to account for the new features such as discontinuity and non-locality. However, Bohm emphasizes that even quantum laws may be abstractions from still more general laws, thus far unknown to us.

This leads him to suggest the term "holomovement" to describe the *totality of movement of unfoldment and enfoldment*, which is taken to include the behavior of both fields known to us today and that of fields still unknown. As we have seen, Bohm thinks that, strictly speaking, the holomovement in a sense comes before fields, and gives rise to the essential properties of fields. *Holomovement* is in this sense a general "metaphysical" or "ontological" concept that goes beyond what is observed today. Bohm's proposal is that the holomovement characterizes physical reality at a fairly fundamental level (while he leaves open the possibility that new ideas might be needed to deal with even deeper aspects of reality that future developments in science may reveal). We pointed out in Chap. 1 that "holomovement" does not mean to imply that there is some single fundamental level of reality where some substantial fields are moving. Instead, the idea is that movement per se is fundamental and gives rise to the essential qualities of fields.

Let us now proceed to consider in more detail what Bohm says about the general architecture or structure of matter in terms of the implicate order.

2.5 The Implicate Order as the General Architecture of Matter

We have seen that in Bohm's implicate order scheme the concept of particle is no longer a primary one. Instead, particles (electrons, protons, etc.) should be seen as relatively stable and recurrent aspects of an underlying movement of unfoldment and enfoldment of various fields. The key point is that such a way of thinking makes it possible, in principle, to account for the central non-mechanistic features of quantum phenomena: discontinuity of movement, wave–particle duality, and non-locality. Also, in this scheme it is possible to derive the phenomena of classical physics as a limiting case. To see how all this works, it is very useful to consider another device, besides the hologram, that helps us both perceive and conceive the nature of the implicate order.

This is an apparently simple device: a tank made of two concentric glass cylinders filled with glycerine, which Bohm saw when watching television. He realized that he could use it as a model to illustrate how the implicate order might constitute the general architecture of matter. It is again important to keep in mind that he uses this device as an analogy which is meant to illustrate the implicate order. The operation of the tank itself can be understood mechanically. However, it serves well to illustrate some features that go beyond the mechanistic order. To see those features, however, it will be necessary to focus on the new holistic features and ignore some mechanical aspects of the device. I shall remind the reader when it is time to "kick away the ladder", so let us move on to consider the analogy.

The device in question consists of two concentric glass cylinders. There is viscous fluid such as glycerine between them. The outer cylinder can be turned very slowly, which results in negligible diffusion of the fluid. Let us place a droplet of insoluble ink in the fluid, and turn the outer cylinder. As we keep turning, the droplet is drawn out into a fine thread-like form that eventually becomes invisible. However, when we turn the outer cylinder in the opposite direction, the thread-like form draws back and suddenly becomes visible, as a droplet essentially the same as the original one (Bohm 1980, p. 179).

To understand the points Bohm wants to make with the device, it is important to consider in more detail how it works (see Fig. 2.1).

Consider first an *element of fluid*, or the glycerine, the stuff in which the ink droplet is placed. When the outer cylinder is turned, the parts at larger radii will move faster than those at smaller radii. Such an element will therefore be deformed, and this explains why it is eventually drawn out into a long thread (Bohm 1980, p. 179).

Consider then the *ink droplet* that is placed in the glycerine.

> This ink droplet consists of an aggregate of carbon particles that are initially suspended in such an element of fluid. As the element is drawn out the ink particles will be carried with it. The set of particles will thus spread out over such a large volume that their density falls below the minimum threshold that is visible. (Bohm 1980, p. 179)

Consider, finally, what happens when the movement is reversed:

> ... as is known from the physical laws governing viscous media... each part of the fluid retraces its path, so that eventually the thread-like fluid element draws back to its original form. As it does so, it carries the ink particles with it, so that eventually they, too, draw together and become dense enough to pass

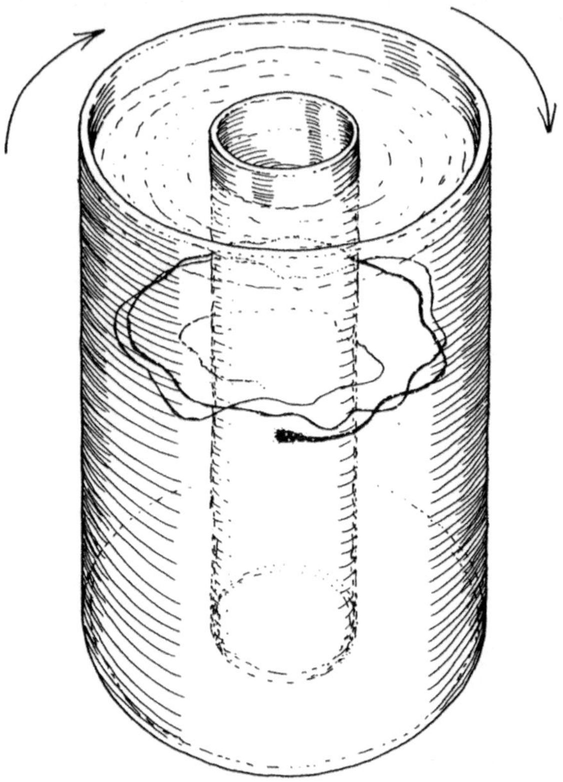

Fig. 2.1. The ink-in-fluid device

> the threshold of perceptibility, so emerging once again as visible droplets. (Bohm 1980, pp. 179–180)

There is a sense in which we can now say that the ink particles, when they have become a long thread, have been *enfolded* into the glycerine, namely in the same sense as we can say that an egg can be folded into a cake.

> Of course, the difference is that the droplet can be unfolded by reversing the motion of the fluid, while there is no way to unfold the egg (this is because the material here undergoes irreversible diffusive mixing). (Bohm 1980, p. 180)

Bohm uses the glycerine tank to illustrate a number of principles. The first one is simply the idea of enfoldment. A localized "unfolded" entity, such as a droplet containing ink particles, can become "enfolded". But because of the reversibility of the situation, it is possible to unfold it again. This makes it reasonable to think of the droplet as defining a *distinguishable ensemble of elements* even in the enfolded state. The "elements" are simply the ink particles constituting the droplet. From the point of view of contemporary metaphysics it is important to note that there is a sense in which the droplet retains its *identifiability* and in an approximate sense even its *identity* in the enfolded state. (For the importance of identifiability to identity in metaphysics see Quine (1960) and Juti (2001, p. 137).) The idea is that there exists a "droplet" (basically, an ensemble of elements or ink particles) in an unfolded state, and (approximately) the *same* droplet can exist in an enfolded state. (I say "approximately", because the Bohmian ontological assumption that movement per se is fundamental only allows one to apply the notion of identity as a useful approximation, rather than as an absolute and literal description of the properties of things).

Of course, the elements of the ensemble in the enfolded state are no longer all in the same region as they were in the unfolded state, as they have spread out through the whole space. But because of the total situation (which involves the possibility of reversibility), we are justified in saying that they remain in a fairly strong sense elements of the original ensemble. Although they can be far apart in space, they are connected. For someone who only saw the ink particles and did not know about the underlying glycerine, the behavior of the ink particles would seem mysterious, especially when, after enfolding them, one starts turning the cylinder backwards. It would seem as if a strange causal necessity were bringing together the ink particles. One of the most important ideas to emerge from Bohm's implicate order scheme is that the fundamental laws of physics involve this type of necessity – but without there being any mechanical medium, such as glycerine, to do the causing!

To better illustrate the sort of causal necessity he has in mind, Bohm asks us to consider two droplets close to each other, and to suppose that the ink particles in one droplet are red while those in the other are blue:

> If the outer cylinder is then turned, each of the two separate elements of fluid in which the ink particles are suspended will be drawn out into a thread-like form, and the two thread-like forms will, while remaining separate and distinct, weave through each other in a complex pattern too fine to be perceptible to the eye (rather like the interference pattern that is recorded on the hologram, which has, however, quite a different origin). The ink particles in each droplet will of course be carried along by the fluid motions, but with each particle remaining in its

> own thread of fluid. Eventually, however, in any region that was large enough to be visible to the eye, red particles from the one droplet and blue particles from the other will be seen to intermingle, apparently at random. (Bohm 1980, p. 180)

In other words, for someone who just saw the ink particles, it would seem that there is a random order in the situation. Now, if one turns the outer cylinder back, the movements of the fluid will be reversed just as in the case of one droplet. As a consequence, one will eventually get back the two droplets. Bohm goes on to use this feature to illustrate how he thinks of the actual physical laws of holomovement, or what he calls "holonomy".

> If one were able to watch what is happening more closely (e.g., with a microscope) one would see red and blue particles that were close to each other beginning to separate, while particles of a given colour that were far from each other would begin to come together. It is almost as if distant particles of a given colour had "known" that they had a common destiny, separate from that of particles of the other colour, to which they were close. (Bohm 1980, p. 180)

The idea of a "common destiny" brings to mind a teleological view of causation. Bohm is, of course, keen to point out that in this case there is no such "destiny", for remember that the glycerine tank is a mechanical analogy which is eventually meant to illustrate something non-mechanical:

> Indeed, we have explained all that has happened mechanically, through the complex movements of the fluid elements in which the ink particles are suspended. But we have to recall here that this device is only an analogy, intended to illustrate a new notion of order. To allow this new notion to stand out clearly, it is necessary to begin by focusing our attention on the ink particles alone, and to set aside the consideration of the fluid in which they are suspended, at least for the moment. (Bohm 1980, pp. 180–181)

Setting aside the consideration of the fluid is like "kicking away the ladder" after reading the *Tractatus* and understanding it. To set aside the fluid and to focus on only the ink particles is to perceive the nature of the laws of the holomovement as Bohm proposes them to be.

As we have already mentioned, one aspect of these laws is that they involve a teleological element. Another is that things can exist in an enfolded state. Two things in an enfolded state can intermingle and yet retain a certain identity:

> When the sets of ink particles from each droplet have been drawn out into an invisible thread, so that particles of both colors intermingle, one can nevertheless say that *as an ensemble* each set is, in a certain way, distinct from the other. (Bohm 1980, p. 181)

This is what was pointed out above and what, in my view, opens up an interesting way of understanding the (approximate) identity of objects. For more generally, we could assume that *any object* is a set or ensemble of elements. Bohm's implicate–explicate scheme suggests that any object can then exist in either explicate (unfolded) mode or implicate (enfolded) mode. Implicate and explicate can then be seen as "ways" or "modes" of being. Implicate resembles the traditional philosophical notion of "potentiality", while explicate resembles "actuality". Remember, however, that the traditional notion of potentiality does not leave room for creativity, as everything that actualizes is already contained in the pre-existing potentiality (see Bunge (1979, p. 206)). As we saw earlier in this chapter, Bohm emphasizes the incompleteness of being, which leaves room for the creation of new potentialities and new modes of unfoldment.

The enfolded mode allows things to intermingle with each other and yet retain an (approximately) separate identity. It makes possible the existence of a high-degree order in a situation which typically would seem to have a very low order.

Of course, one has to be careful in assuming that any object is reducible to a set or ensemble of elements. What are such "elements" anyway? In the analogy they are ink particles. But this shows one of the limits of the analogy if the purpose is to illustrate a new notion of order. In physics we are more fundamentally dealing with fields, and thus the "elements" have to be thought of as parts of a field. And even more fundamentally, as we have already mentioned, because of relativistic considerations, even fields should not be taken in a substantial sense to involve some physical medium (see Bohm and Hiley (1993, pp. 356–7)). For example, the notion of "ether" in classical physics was given up as a result of the Michelson–Morley experiment, and the electromagnetic fields are thought to exist in themselves. Quantum field theoretical considerations do introduce the notion of the vacuum as being full of energy (and thus not empty), but the vacuum should not be understood in too substantial terms. As I have emphasized before, Bohm thinks it is better to think of "movement" as the fundamental concept.

It is "movement" (indeed, the "holomovement") that is fundamental, without there being something that is doing the moving (either particles or fields). Particles and fields are then aspects that emerge from the movement, which is primary. Thus, although the ink-in-fluid analogy makes reference to elements, and ensembles of elements, when we apply the analogy in a quantum context, we have to be careful not to interpret "elements" in too substantial terms (for example, as particles or fields).

Another thing to consider is that the emphasis on "ensembles of elements" would seem to emphasize *set theory* as fundamental to Bohm's scheme. This again would not be very radical or holistic, for set theory seems to be a logical reflection of the mechanistic or explicate order. Bohm has elsewhere commented on the relevance of set theory in the context of his new world view, which gives a primary role to "undivided wholeness in movement". His view implies that

> a new kind of mathematics has to be developed, to fit in with this world view. Thus, current mathematics is based largely on set theory, which is in turn an articulation of the traditional world view, in which the whole of existence is regarded as constituted out of a collection of separately existent things. What is now needed is a mathematics in which the primary function of symbols is to call attention directly to aspects of a whole movement. (Bohm 1976, p. 47)

Accordingly, the role of mathematics in the implicate order framework is to draw attention to the processes of unfoldment and enfoldment, rather than to the elements that are unfolding and enfolding. Thus, although the notion of an element is very useful in Bohm's ink droplet analogy, one has to be careful not to think of the elements as fundamental in the sense of set theory. Instead, a mathematics which describes movement is needed. For Bohm's proposals for such mathematics (which involve various kinds of algebra), see Bohm (1980, pp. 157–171) and Bohm and Hiley (1993, pp. 363–378).

Let us return to our discussion of the ink-in-fluid analogy. There is a possibility of distinguishing between ensembles of elements. However, this distinguishability is a fairly subtle property. It involves the total situation and depends on the operation of what Bohm calls an "inherent overall necessity":

> This distinction [between ensembles] is not in general evident to the senses, but it has a certain relationship to the total situation out of which the ensembles have come. This situation includes

> the glass cylinders, the viscous fluid and its movements, and the original distribution of ink particles. It may then be said that each ink particle belongs to a certain distinct ensemble and that it is bound up with the others in this ensemble by the force of an overall necessity, inherent in this total situation, which can bring the whole set to a common end (i.e., to reconstitute the form of a droplet). (Bohm 1980, p. 181)

This brings out another important feature of Bohm's idea of the implicate order. An implicate order in a given situation is not necessarily directly perceivable. It typically depends on the total context which makes possible the existence of the implicate order (i.e. the enfolded state), and also the unfoldment of an implicate order into an explicate order. Further, as we have already indicated, there is assumed to be a factor or *force* which brings elements that form a given implicate order to constitute an explicate order. Bohm denotes that factor the "overall necessity". What is important is that such "overall necessity" is *inherent* in the overall situation – which means that we cannot understand its origin without going outside of the situation.

This is a useful idea when one tries to think of the nature of physical laws, and of laws more generally (for example, in the biological and psychological domain). The suggestion is that in a given context it is likely that there appear some regularities, the origin of which cannot be understood in the context in question, but which just seem to be "fundamental" and "absolute". For example, the laws of Newtonian physics appeared to have such a quality. However, when one manages to go outside of the context in question, new phenomena may appear in which the old regularities no longer apply. Thus, quantum phenomena suggest that "particles" have wave properties, giving rise to particle interference, which is completely against the predictions of Newton's laws. The "absolute" laws of Newton suddenly appear as approximations that only apply in a given limited context and give completely wrong predictions in a broader context. But the same could happen to quantum laws, should we one day be able to go outside the current context of physics. Thus, physical laws typically appear to have a certain absolute quality, in the sense that there seems to be no obvious reason within the context in which they hold that would explain why they hold. At the same time, there seems to be no reason to assume that a given law really is absolute, holding in all possible contexts (cf. Bohm (1957), Cartwright (1983), and Earman et al. (2002)).

Bohm reminds us again that, when considering the glycerine tank, we are dealing with a mechanical analogy of the overall necessity rather than the "real thing":

> In the case of this device, the overall necessity operates mechanically as the movement of fluid, according to certain well-known laws of hydrodynamics. ... however, we will eventually drop this mechanical analogy and go on to consider the holomovement. In the holomovement there is still an overall necessity (... "holonomy") but its laws are no longer mechanical. Rather... its laws will be in a first approximation those of the quantum theory, while more accurately they will go beyond even these, in ways that are at present only vaguely discernible. (Bohm 1980, p. 181)

The above begins to illustrate further how Bohm thinks of the deeper nature of causality in his implicate order framework. The mechanistic view of causality emphasizes interaction via direct push and pull (as in the Cartesian concept of matter), or else interaction in terms of forces between particles, forces which diminish with distance (e.g. the gravitational and electromagnetic forces of classical physics). In the context of the implicate order, however, Bohm wants to give up this kind of mechanical causality as fundamental, and replace it with a notion of causality which, as we have mentioned, clearly has *teleological* features:

> Nevertheless, certain similar principles of distinction will prevail in the holomovement as in the analogy of the device made up of glass cylinders. That is to say, ensembles of elements which intermingle or inter-penetrate in space can nevertheless be distinguished, but only in the context of certain total situations, in which the members of each ensemble are related through the force of an overall necessity inherent in these situations, that can bring them together in a specifiable way. (Bohm 1980, p. 181)

The key point is that in the implicate order framework the fundamental laws are not mechanical interactions between particles in the explicate order; rather the fundamental laws have to do with the way ensembles of elements ("droplets") unfold from the implicate order to the explicate order. This in turn depends on the nature of the total situation, as well as on the mutual relationships of the ensembles that are in the enfolded state.

We will come back to consider the causal architecture of the holomovement later. For now, let us follow Bohm in developing some further concepts which will also be useful when we discuss the nature of consciousness in the next chapter.

Having explained what he means by "distinguishable ensembles", Bohm next wants to put them into an order. In this way he hopes to use the ink-in-fluid model to illustrate the notion of implicate order even better. According to him the simplest notion of order is that of a sequence or succession. Think of building a house from bricks, and beginning to construct a wall by first putting bricks next to each other in a straight line. Or think of the series of natural numbers 1, 2, 3, ... These "similar differences" forming a sequence exemplify the simplest cases of order. His strategy is to start with such simple cases of order and develop more complex notions out of these.

First of all, he uses the device to enfold a simple *pre-existent explicate order.* We can do this by inserting into the fluid a large number of droplets, set close to each other and arranged in a line. Let's label them A, B, C, D, ... Let's then turn the outer cylinder many times, so that each of the droplets gives rise to an ensemble of ink particles, enfolded in so large a region of space that particles from all droplets intermingle. Let's label the successive ensembles A′, B′, C′, D′, ... We have now enfolded an entire *linear order* into the fluid (Bohm 1980, p. 182).

A much more exciting order is something Bohm calls an *intrinsically implicate order.* This is an implicate order which cannot be made all explicate at once. It is straightforward to illustrate this with the device. Let us insert an ink droplet, A, and turn the outer cylinder n times. Then we insert a second ink droplet, B, at the same place and again turn the cylinder n times. And then we repeat the procedure with further droplets C, D, E, ... The resulting ensembles of ink particles a, b, c, d, e, ... will now differ in a new way:

> ... when the motion of the fluid is reversed, the ensembles will successively come together to form droplets in an order opposite to the one in which they were put in. For example, at a certain stage the particles of ensemble d will come together (after which they will be drawn out into a thread again). This will happen to those of c, then to b, etc. It is clear from this that ensemble d is related to c as c is to b, and so on. (Bohm 1980, pp. 182–183)

But this is not a transformation of a linear order in space (as was the transformation of the sequence A′, B′, C′, D′, ...), for in general only *one* of these ensembles will unfold at a time; when any one is unfolded, the rest are still enfolded.

> In short, we have an order which cannot all be made explicate at once and which is nevertheless real, as may be revealed when successive droplets become visible as the cylinder is turned. (Bohm 1980, p. 183)

Bohm calls this an intrinsically implicate order, to distinguish it from an order that may be enfolded but that can unfold all at once into a single explicate order. In the context of this kind of intrinsically implicate order, an explicate order is a particular case of a more general set of implicate orders.

Bohm's next step is to combine both of the above-described types of order. Let us insert a droplet, A, in a certain position and turn the cylinder n times. Then we insert a droplet, B, in a slightly different position and turn the cylinder n more times (so that A has been enfolded by $2n$ turns). And then we insert C along the line AB and turn the cylinder n more times, so that A has been enfolded by $3n$ turns, B by $2n$ turns, and C by n turns. We proceed in this way to enfold a large number of droplets. Then we move the cylinder fairly rapidly in the reverse direction.

> If the rate of emergence of droplets is faster than the minimum time of resolution of the human eye, what we will see is apparently a particle moving continuously and crossing the space. (Bohm 1980, p. 183)

Bohm now suggests that such enfoldment and unfoldment in the implicate order provides a new model of, for example, an electron. Such a model is quite different from that provided by the current mechanistic notion of a particle that exists at each moment only in a small region of space and changes its position continuously with space. The essential point about the new model is that the electron is instead to be understood through a *total set of enfolded ensembles*, which are generally *not localized in space.* At any given moment, one of these may be unfolded and therefore localized, but in the next moment, this one enfolds to be replaced by the one that follows (Bohm 1980, p. 183).

It is important to realize that one is here using *many* "droplets" or "ensembles of ink particles" to model a *single* electron. In physics, of course, one does not take the ink-in-fluid model as fundamental. What is fundamental is instead the notion of a field, and (in Bohm's view) even more fundamentally, just movement. But if we translate Bohm's ink-in-fluid model of the electron to the more physically accurate model in terms of fields, we say that the single electron can be understood through a *total set of fields* which are generally not localized in space.

These fields involve a constant movement of enfoldment and unfoldment of waves, as described above. At any given moment, one of these waves may have all of its amplitude localized in a single region (and in this sense be unfolded to this region), to produce a particle-like manifestation. In the next moment, this wave spreads out or enfolds, to be replaced by the next "incoming wave" (which is either a new wave, or else the outgoing wave that is somehow reflected back; it seems that Bohm in this 1980 text assumes that it is a new wave).

We are used to the idea of the continuity of existence of particles. It is very convenient to think of an electron just "sitting there" in a region of space and continuing to exist over time, in the same manner as macroscopic physical objects seem to exist. However, the experimental evidence from relativity and quantum physics strongly suggests that the notion of this kind of continuously existing particle cannot be the most fundamental notion in physics (although it can of course be extremely convenient in many situations). Thus, a more fundamental physical theory (that is, say, valid at the temporal level of the Planck time 10^{-43} s) has to replace the notion of the continuity of existence of a particle by the notion of very rapid recurrence of similar forms, changing in a simple and regular way (see Bohm and Hiley (1993, pp. 367–368)). The idea is that such recurrence approximates the notion of continuity of existence of particle, though, strictly speaking, what is actually taking place is a rapid recurrence of unfolding-enfolding activity. Bohm suggests that what happens here is in some ways similar to what happens with a rapidly spinning bicycle wheel, which gives the impression of a solid disk, rather than of a sequence of rotating spokes.

When we think of the "moving droplet" in the glycerine tank, the notion of a continuously existing particle is useful, but more fundamentally this "particle" is only an abstraction that is manifest to our senses. The "particle" is something *we* abstract from the more fundamental movement of unfoldment and enfoldment. What the electron (when understood through the ink-in-fluid model) actually is, is

> always a totality of ensembles, all present together, in an orderly series of stages of enfoldment and unfoldment, which intermingle and inter-penetrate each other in principle throughout the whole of space. (Bohm 1980, p. 184)

We could summarize this by saying that the implicate order typically involves the *co-presence of elements* (or ensembles of elements) *at different degrees of enfoldment*. If you like, this is one of the "marks" of an implicate order. This idea will be relevant later on when we discuss time consciousness in terms of the implicate order.

If one seeks for a more physically accurate model of an electron, one could say that the electron is a totality of fields/waves, all present together, in an orderly series of stages of enfoldment and unfoldment, which intermingle and interpenetrate each other, in principle, throughout the whole of space. In this sense a single electron is "enfolded" into the whole universe, roughly in an analogous sense that an egg can be folded into the whole cake!

We have seen above how a new model of a single electron can be provided in the context of the ink-in-fluid device. It is straightforward to extend this and model many electrons, as Bohm points out:

> It is further evident that we could have enfolded any number of such "electrons", whose forms would have intermingled and inter-penetrated in the implicate order. Nevertheless, as these forms unfolded and became manifest to our senses, they would have come out as a set of "particles" clearly separated from each other. (Bohm 1980, p. 184)

This provides a very clear illustration of how the "explicate order" of separate particles arises from the implicate order. Bohm next points out that we could use the ink-in-fluid device to provide a model of physical phenomena covered by classical physics:

> The arrangement of ensembles could have been such that these particle-like manifestations came out "moving" independently in straight lines, or equally along curved paths that were mutually related and dependent, as if there had been a force of interaction between them. Since classical physics traditionally aims to explain everything in terms of interacting systems of particles, it is clear that in principle one could equally well treat the entire domain that is correctly covered by such classical concepts in terms of our model of ordered sequences of enfolding and unfolding ensembles. (Bohm 1980, p. 184)

If Bohm's aim is to provide us with a new concept of the general structure or architecture of matter, it is of course important that he can model the domain of classical physics. A proposal for an architecture of matter that cannot do justice to the domain of classical physics cannot take us very far. The implicate order framework thus provides an alternative (at least conceptually and in principle) to Newtonian mechanics or the Hamilton–Jacobi theory in the domain of classical physics. The crucial point, however, is that the new model can account for the puzzling quantum phenomena:

> What we are proposing here is that in the quantum domain this model is a great deal better than is the classical notion of an interacting set of particles. Thus, although successive localized manifestations of an electron, for example, may be very close to each other, so that they approximate a continuous track, this need not always be so. In principle, discontinuities may be allowed in the manifest tracks – and these may, of course, provide the basis of an explanation of how... an electron can go from one state to another without passing through states in between. This is possible, of course, because the "particle" is only an abstraction of a much greater totality of structure. This abstraction is what is manifest to our senses (or instruments) but evidently there is no reason why it has to have continuous movement or indeed continuous existence. (Bohm 1980, p. 184)

Thus, the ink-in-fluid model of the electron provides an elegant account of the famous "quantum jumps". Indeed, the quantum jumps provide evidence for the ink-in-fluid model of an electron, or other such models that can easily accommodate discontinuity of motion. In contrast, models which presuppose the continuous existence and movement of particles have serious difficulties in accommodating quantum jumps.

Similarly, one can model wave-particle duality with the ink-in-fluid analogy. It is characteristic of quantum phenomena that "if the total context of the process is changed, entirely new modes of manifestation may arise." In other words, electrons can show different properties (e.g. particle-like, wave-like, or something in between), depending on the environmental context within which they exist and are observed. In terms of the ink-in-fluid analogy, Bohm notes that if the cylinders are changed, or if obstacles are placed in the fluid, the form and order of manifestation will be different. He emphasizes that such dependence – the dependence of what manifests to observation on the total situation – has a close parallel to the wave–particle duality of quantum phenomena.

For the above kinds of reasons, Bohm suggests that the implicate order generally gives a much more coherent account of the quantum properties of matter than does the traditional mechanistic order. His proposal thus is that *the implicate order ought to be taken as fundamental.* This proposal brings to focus the tension between the traditional mechanistic approach and the implicate order approach. We could say that the mechanistic approach is based on the assumption that the explicate order is the fundamental basis of existence. Bohm emphasizes that even in terms of the mechanistic approach

> it may of course be admitted that in a certain sense at least, enfoldment and unfoldment can take place in various specific situations (e.g., such as that which happens with the ink droplet). However, this sort of situation is not regarded as having a fundamental kind of significance. All that is primary, independently existent, and universal is thought to be expressible in an explicate order, in terms of elements that are externally related (and these are usually thought to be particles, or fields, or some combination of the two). Whenever enfoldment and unfoldment are found actually to take place, it is therefore assumed that these can ultimately be explained in terms of an underlying explicate order through a deeper mechanical analysis (as, indeed, does happen with the ink-droplet device). (Bohm 1980, p. 185)

So the issue is not whether enfoldment and unfoldment take place in various specific situations. They obviously do. The issue rather is what is the *status* of enfoldment and unfoldment. For the mechanist, enfoldment and unfoldment are secondary, derivative features, reducible to activities of the particles and/or fields in the explicate order, which is thought to be fundamental. For the advocate of the implicate order approach, enfoldment and unfoldment are primary, and the explicate order – including elementary fields and particles – can be derived from these. Mechanical analysis has been very successful in physics, but Bohm proposes that as we consider the deeper, more fundamental levels of physical existence, mechanical analysis no longer works. To deal with these levels, another kind of description is needed, a description which assumes that enfoldment and unfoldment are primary.

There is, however, paradoxically a sense in which Bohm's ink-droplet device can play into the hands of the mechanist. For it provides a beautiful illustration of how fairly subtle enfolding–unfolding processes can be reduced to a mechanistic basis. This is why it is extremely important, if one wants to understand Bohm's proposal, to understand the limitations of the ink-droplet device, and to "throw it away" once one has understood the principles of the implicate order that it can illustrate so well (such as enfoldment and unfoldment, the idea of an electron as a set of co-present ensembles at different degrees of enfoldment, the way the continuous existence of a particle can be approximated by rapid recurrence of similar forms, etc.). Bohm's suggestion is that what is happening at the more fundamental levels of physical existence is in some ways similar to what happens with the ink-droplet device. But there are differences. In particular, as we have already mentioned, in a more accurate physical theory we are to think of the electron as a set of fields which are in a movement of unfoldment and enfoldment. And the further idea is that the essential properties of the fields themselves arise from movement. It is movement as unfoldment and enfoldment that is fundamental. He writes:

> Our proposal to start with the implicate order as basic, then, means that what is primary, independently existent, and universal has to be expressed in terms of the implicate order. So we are suggesting that it is the implicate order that is autonomously active while ... the explicate order flows out of a law of the implicate order, so that it is secondary, derivative, and appropriate only in certain limited contexts. (Bohm 1980, p. 185)

Bohm's use of the terms "primary", "independently existent", "universal", and "autonomously active" suggests that he is thinking of that part of reality where the implicate order prevails as something similar to what philosophers have traditionally called "substance". Of course, Bohm's substance is movement, so it differs from, for example, Aristotle's or Descartes' notion of substance. However, the key point is that for Bohm there is only one fundamental substance, the movement in which the implicate order prevails, and the explicate order does not have the status of a substance, as it is "secondary" and "derivative".

Let us again consider how Bohm thinks about the nature of laws in his ontology:

> the relationships constituting the fundamental law are between the enfolded structures that interweave and inter-penetrate each other, throughout the whole of space, rather than between the abstracted and separated forms that are manifest to the senses (and to our instruments). (Bohm 1980, p. 185)

As we have already mentioned above, this suggests an alternative way of thinking about the nature of physical law. The traditional view is to think that the fundamental laws of physics are forces between separate elements in the explicate order. It is also sometimes assumed that such forces are mediated by yet other particles, which further underlines the mechanistic character of some of the current ways of thinking about the nature of fundamental law. Bohm proposes an alternative view. Instead of focusing on the relationships of unfolded elements in the explicate order (particles and/or fields), we ought to be focusing on the relationships between enfolded structures in the implicate order. The idea is that what happens in the observable arena of the explicate order is determined by relationships at the level of the implicate order. Physics is concerned with explaining regularities in nature in terms of laws, and Bohm urges that when in search of the more fundamental laws, physics ought to turn its attention to the level of the implicate order, instead of desperately trying to explain all regularities in terms of mechanistic interaction in the explicate order.

Consider also a later quotation from Bohm and Hiley, which further illustrates some of the above points:

> These notions of enfoldment and unfoldment can be understood in terms of the Cartesian order as being only particular cases of movements of fields not having the kind of general necessity that we would attribute, for example, to laws of nature. Rather what is significant for such laws of physics is considered to be the order of separate points. What we are proposing here is to turn this notion upside down and say that the implicate order will have the kind of general necessity that is suitable for expressing the basic laws of physics, while the explicate order will be important within this approach only as a particular case of the general order. (Bohm & Hiley 1993, p. 354)

The tradition in physics has been to take the "order of separate points" as fundamental, and to express all laws of nature in terms of this order. Bohm and Hiley suggest that we instead have to take the implicate order as fundamental, and express the basic laws of physics in terms of this order. "The order of separate points" (the explicate order) can then be derived as a special case of the more "general order" (the implicate order). The point is that the implicate order has "general necessity" – it draws our attention to the fundamental laws of nature.

One good example of a physical phenomenon which is difficult to understand in terms of the explicate order is quantum non-locality, and it is now time to consider this.

2.6 Non-locality and the Implicate Order

Bohm notes that he has thus far presented the implicate order as a process of enfoldment and unfoldment taking place in ordinary three-dimensional space (Bohm 1980, p. 186). However, he goes on to suggest that quantum theory implies the need to extend this process to a multidimensional space. The reason is the famous non-local relationship of quantum theory, which Einstein, Podolsky, and Rosen brought to focus in their 1935 paper (it is also known as the EPR relation, and because the phenomenon is so puzzling it has been called the "EPR paradox"). Bohm characterizes this relationship as "a non-causal connection of elements that are distant from each other". What is a "non-causal connection"?

The basic idea in the EPR experiments is that systems A and B (usually photons, but also other quantum systems such as atoms, electrons, etc.) have first been together, but have then been separated from each other (in current experiments the maximum distance between A and B is ca 50 km). Interacting with A then seems to influence B instantaneously, or at least much faster than the speed of light (the maximum speed for transmission of signals according to relativity) allows. This is mysterious.

The physicist Stig Stenholm once presented in a Finnish television program some useful analogies that help to understand quantum non-locality. There are two analogies: the first illustrates a kind of non-locality that is familiar from our everyday experience, while the second illustrates the more exotic quantum non-locality. Consider a situation in which two people, let's say you and I, meet in Stockholm. We have a coin. I split it in the middle into two halves, so that one of them has "heads" and the other one "tails", but neither of us looks to see which half has which. I keep one half and you go to Paris with the other half. When I then look at my half of the coin and see, say, "heads", I immediately know that you will see "tails" when you look at your half (assuming that nothing has happened to your half of the coin, you haven't had too much Beaujelois to drink, etc.). There is an instantaneous change in my knowledge about things far away, and no signals need to be transmitted. This is everyday experience, and involves no need to change the laws of physics.

However, to understand the non-locality of quantum theory, it is useful to consider another analogy. Think of a situation in which we again meet in Stockholm and now have two dice. You take one and go to Paris; I keep the other and stay in Stockholm. When I throw my die, and get, say "three", I of course cannot predict from that result what you will get when you throw your die in Paris. In the everyday world the dice are independent of each other. In order to illustrate the nature of quantum non-locality, let us now assume that the dice were behaving like photons or electrons, i.e. obeying the laws of quantum theory. Again, we meet in Stockholm, and you take one of the dice and go to Paris. Now, if the dice were following quantum laws, the following would happen. When I throw my die and get, say, "three", I could immediately tell which result (say, "six") you will get when you throw your die in Paris. And I will be able to predict your result even if you were to throw your dice immediately after I throw mine (for example, you can throw your die before a light signal has managed to travel from Stockholm to Paris, and I can still predict your result correctly). This is a simplified analogy to real physics experiments, but it serves well to bring out the meaning of quantum non-locality. There is clearly a connection between the results of the "throws", but there seems to be no explanation of the connection in terms of the kinds of causal processes that physics traditionally allows for (e.g. direct push and pull, influence via electromagnetic or gravitational forces). This is why Bohm called it "a non-causal connection".

For him the key point is that

> the analysis of a total system into a set of independently existent but interacting particles breaks down in a radically new way. One discovers, instead, both from consideration of the meaning of the mathematical equations and from the results of the actual experiments, that the various particles have to be taken literally as projections of a higher-dimensional reality which cannot be accounted for in terms of any force of interaction between them. (Bohm 1980, pp. 186–7)

To illustrate this notion of projection, Bohm considers another analogy. Imagine a rectangular tank in which there is a fish swimming. Assume further that we are filming the tank with two video cameras A and B, one directed to a side of the tank and the other to one of the ends. We then attach one monitor to each camera. When we look at the monitors, we typically see two correlated images. This serves to illustrate the idea of correlated projections, which is very important for Bohm:

> Of course, we know that the two images do not refer to independently existent though interacting actualities ([so that] one image could be said to "cause" related changes in the other). Rather, they refer to a single actuality, which is the common ground of both (and this explains the correlation of images without the assumption that they causally affect each other). This actuality is of higher dimensionality than are the separate images on the screen. To put it differently: the images on the screens are two-dimensional *projections* (or facets) of a three-dimensional reality. In some sense this three-dimensional reality holds these two-dimensional projections within it. Yet, since these projections exist only as abstractions, the three-dimensional reality *is* neither of these, but rather it is some-

> thing else, something of a nature beyond both. (Bohm 1980, pp. 187–8)

This opens up a way to think about quantum non-locality. Bohm proposes that we likewise regard each of the "particles" constituting a system as a projection of a "higher-dimensional" reality, rather than as a separate particle, existing together with all the others in a common three-dimensional space (Bohm 1980, p. 188). Thus, for example, when we are considering two systems (e.g. atoms) that are non-locally correlated, we ought to see them as three-dimensional projections of a six-dimensional reality. We will see later how Bohm suggests that the relationship of mind and body is analogous to the relationship between the systems in an EPR situation. Mind and body are to be seen as lower-dimensional projections of a higher-dimensional reality.

Bohm concludes that quite generally the implicate order has to be regarded as a process of enfoldment and unfoldment in a higher-dimensional space. The processes of enfoldment and unfoldment in three dimensions (as with the ink-in-fluid analogy and the hologram) are simplifications that only apply under certain conditions. For example, strictly speaking, the electromagnetic field (which underlies holography) obeys quantum laws and is thus a multidimensional reality.

> Quite generally, then, the implicate order has to be extended into a multidimensional reality. In principle this reality is one unbroken whole, including the entire universe with all its "fields" and "particles". Thus we have to say that the holomovement enfolds and unfolds in a multidimensional order, the dimensionality of which is effectively infinite. (Bohm 1980, p. 189)

Of course, this raises the question of how to reconcile the apparent fact of our ordinary three-dimensional reality with the multidimensional reality postulated by Bohm. As we have seen, he assumes that it is a basic feature of the holomovement that it gives rise to the relatively autonomous sub-order that we meet in everyday sensory experience and which he labels the "explicate order". Indeed, the possibility of getting a stable explicate order out of the flux of the holomovement is so important for Bohm that he elevates it into a principle, which he calls "the principle of relative autonomy of sub-totalities".

2.7 Cosmology and the Implicate Order

Bohm proposes the implicate order as a framework in which to discuss the architecture of matter, and hopes to extend it to biological and psychological phenomena, thus making it into a proposal about the architecture of existence as a whole. One important issue here is cosmology, or our views about the total order of the universe at large. The implicate order might well help us to discuss such exotic quantum features as the discontinuity of motion and non-locality; but does it have anything illuminating to say about cosmology? This is what we will consider in this section.

Bohm discusses his views about cosmology in some detail (Bohm 1980, pp. 190–3), but we will here consider only some main aspects his discussion. He first draws attention to the fact that quantum field theory implies that empty space is not at all empty, but on the contrary contains huge amounts of energy. This is because of the so-called zero-point energy, which he estimates as follows:

> If one computes the amount of energy that would be in one cubic centimetre of space, with this shortest possible wavelength [10^{-33} cm], it turns out to be very far beyond the total energy of all the matter in the known universe. (Bohm 1980, pp. 190–1)

This leads him to suggest that what we call empty space contains an immense background of energy (for further discussion, see, for example, Hiley (1991) and Saunders (2002)). This in turn leads to a radically new perspective on ordinary matter:

> ... matter as we know it is a small, "quantized" wavelike excitation on top of this background, rather like a tiny ripple on a vast sea ... this vast sea may play a key part in the understanding of the cosmos as a whole. (Bohm 1980, p. 191)

Bohm had already suggested this type of calculation in his 1957 book *Causality and Chance in Modern Physics* (Bohm 1957, pp. 163–4), and it seems to me that recent developments in "mainstream cosmology" are moving in the same direction (see, for example, the proposal by the Quantum Universe Committee (2005)). He next considers the idea of space containing a huge energy in relation to earlier ideas about whether space is full or empty, and notes that while modern science has generally favoured the Democritean idea of space as empty, during the 19th century the Parmenidian–Zenoan idea of space as full was entertained through the ether hypothesis. Matter was thought to be like ripples or vortices in the ether, transmitted through the plenum as if it were empty (Bohm 1980, p. 191). He then notes that a similar notion is used in modern physics, with some very interesting consequences:

> According to the quantum theory, a crystal at absolute zero allows electrons to pass through it without scattering. They go

> through as if the space were empty. If the temperature is raised, inhomogeneities appear, and these scatter electrons. If one were to use such electrons to observe the crystal (i.e. by focusing them with an electron lens to make an image) what would be visible would be just the inhomogeneities. It would then appear that the inhomogeneities exist independently and that the main body of the crystal was sheer nothingness. (Bohm 1980, p. 191)

Bohm then suggests that there is an analogy between the situation with the crystal and our everyday perception. Common experience suggests that we are perceiving an empty space through our senses, but Bohm proposes, radically, that we are in fact perceiving the plenum, which is the ground for the existence of everything, including ourselves. Just as the electrons only "see" the inhomogeneities in the crystal and not the crystal itself, in an analogous way the idea is that in sensory perception we only see the "inhomogeneities" in the plenum, which appear as the familiar physical objects of everyday experience. This is a very radical idea, as it suggests that everyday perception is in fact a much more exotic process than we customarily take it to be. Note also that the brain (and body) itself consists of ordinary matter and thus of the same kind of' "inhomogeneities" in the plenum. Perception thus involves a relation between two distinct "inhomogeneities" in the plenum, the object (e.g. a chair) and the brain/body. Somehow out of this relation emerges the "virtual reality", the perceptual world we consciously experience.

In the light of modern physics it thus seems that space is full rather than empty. There is, however, a difference between Bohm's proposal and the 19th-century notion of the ether:

> The plenum is... no longer to be conceived through the idea of a simple material medium, such as an ether, which would be regarded as existing and moving only in a three-dimensional space. Rather, one is to begin with the holomovement, in which there is the immense "sea" of energy... This sea is to be understood in terms of a multidimensional implicate order... while the entire universe of matter as we generally observe it is to be treated as a comparatively small pattern of excitation. This excitation pattern is relatively autonomous and gives rise to approximately recurrent, stable and separable projections into a three-dimensional explicate order of manifestation, which is more or less equivalent to that of space as we commonly experience it. (Bohm 1980, p. 192)

Such an idea is, of course, very interesting for our concept of reality. It also presents a challenge for metaphysics, the branch of philosophy which was neglected in much of the 20th century but began a comeback toward the end of the century. Metaphysics deals with "the most basic and general features of reality and our place in it" (Kim & Sosa 1999, p. ix). Now, surely what Bohm is describing above is fairly basic and general. Note especially that he is not taking three-dimensional space as a fundamental feature of reality but rather as a derivative feature, something that arises as a result of a projection from the holomovement, in which a multidimensional implicate order prevails. However, when one looks at recent work in metaphysics, the notion of space typically plays a very fundamental role. For example, spatial location is thought to be a crucial feature which can be used to identify objects and distinguish abstract objects (which lack a spatial location) from concrete or physical objects (which typically have such a location). Bohm, and modern physics more generally, suggests, however, that there are other features which might be more fundamental than spatial location for metaphysics to consider. Taking such ideas arising from the new physics seriously might thus open up entirely new possibilities and questions for metaphysicians. Of course, some philosophers are already engaged in this, but perhaps most metaphysicians still do their research without considering the quantum and relativistic revolutions (see, however, Callender and Huggett (2001), for instance).

Bohm next considers the well-known idea of a "big bang" in the light of the above, noting that in his approach, "this 'big bang' is to be regarded as actually just a 'little ripple'". He gives an analogy:

> ... in the middle of the actual ocean (i.e., on the surface of the Earth) myriads of small waves occasionally come together fortuitously with such phase relationships that they end up in a certain small region of space, suddenly to produce a very high wave which just appears as if from nowhere and out of nothing. Perhaps something like this could happen in the immense ocean of cosmic energy, creating a sudden wave pulse, from which our "universe" would be born. This pulse would explode outward and break up into smaller ripples that spread yet further outward to constitute our "expanding universe". The latter would have its "space" enfolded within it as a special distinguished explicate and manifest order. (Bohm 1980, p. 192)

This is an attempt to answer the question about what there was before the big bang. Bohm's suggestion certainly puts the big bang into perspective! He also reminds us of the possible limitation of the current idea (at least in 1980) that the universe can be understood as if it were self-existent and independent of the sea of cosmic energy. It all depends upon how far the notion of a relatively independent sub-totality applies to the universe. He notes that black holes, for example, may lead us into an area in which the cosmic background of energy is important. He also points out that his above suggestions imply that there may be many other such expanding universes. It might be interesting to compare Bohms ideas with some recent research on the big bang; see, for example, Steinhardt and Turok (2006).

Another point to remember is that even this vast sea of cosmic energy takes into account only what happens on a scale larger than the critical length of 10^{-33} cm.

> But this length is only a certain kind of limit on the applicability of ordinary notions of space and time. To suppose that there is nothing beyond this limit at all would indeed be quite arbitrary. Rather, it is very probable that beyond it lies a further domain, or set of domains, of the nature of which we have as yet little or no idea. (Bohm 1980, p. 193).

Bohm has provided us with a *progression* from the explicate order to a simple three-dimensional implicate order, then to a multidimensional implicate order, then to an extension of this to the immense "sea" in what is sensed as empty space. He notes that the next stage may well lead to yet further enrichment and extension of the notion of implicate order, beyond the critical limit of 10^{-33} cm, or it may lead to some basically new notions which could not be comprehended even within the possible further developments of the implicate order. This is summarized in Fig. 2.2.

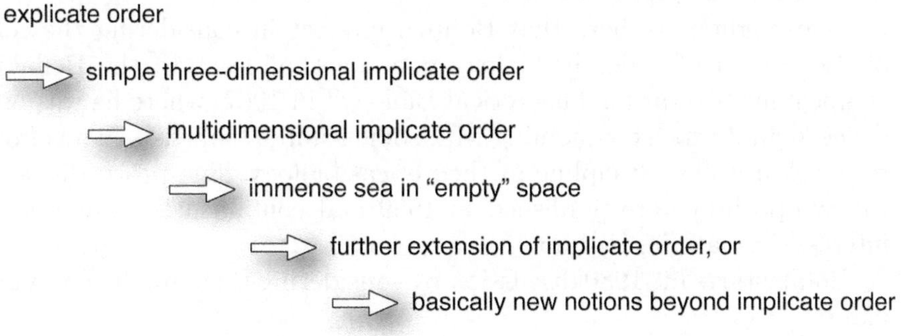

Fig. 2.2. The progression from the explicate to the implicate order

The fact that we may one day need to give up the idea of the implicate order as fundamental does not, however, mean that it would be completely useless. As we have seen, Bohm thinks that theories and ideas typically have their own proper domains in which they provide a coherent view. The implicate order is meant as a very general view, trying to describe the general architecture of existence. However, even this description may turn out to be limited if one, say, managed to go beyond the critical limit of 10^{-33} cm. Bohm is thus open to the idea that he has not arrived even at an outline of an absolute and final truth. However, he thinks that it is characteristic of the holomovement that each sub-totality has a relative autonomy which makes it possible to study it in its own right, up to a point. Thus, although it is important to remember that something completely different is needed if we manage to go beyond the immense energies of empty space, there is much to study in the sub-totalities that have been found thus far (Bohm 1980, p. 193).

2.8 Extending the Implicate Order to Biological Phenomena

I have mentioned several times that although Bohm initially developed the implicate order as a framework in which one could hope to unite quantum theory and relativity in a coherent way, his ambition was to extend it to biological and psychological phenomena, thus making it into a proposal about the architecture of existence in general, rather than just about the architecture of inanimate matter, the subject matter of physics. In this section I will briefly consider the extension to biological phenomena, as he presented it in the last chapter of *Wholeness and the Implicate Order*.

One might note here that Bohm's interest in considering the role of the notion of order in biology goes back at least to the Bellagio symposium "Toward a Theoretical Biology" in 1967, where he was one of the keynote speakers, joining with some leading biologists in an effort to develop a new discipline of theoretical biology. The proceedings of the symposium were published in 1969 and contain a fair amount of interesting material by him.

Bohm starts his 1980 discussion by considering the growth of a living plant:

> This growth starts from a seed, but the seed contributes little or nothing to the actual material substance of the plant or to the

> energy needed to make it grow. This latter comes almost entirely from the soil, the water, the air and the sunlight. According to modern theories the seed contains *information*, in the form of DNA, and this information somehow "directs" the environment to form a corresponding plant. (Bohm 1980, pp. 193–4)

This idea is, of course, very close to the idea of active information, which, as we saw in Chap. 1, plays a key role in the ontological interpretation of quantum theory (Bohm 1990; Bohm & Hiley 1987, 1993). Here he is primarily concerned with the similarity of the situation to the ink-in-fluid model of the electron:

> In terms of the implicate order, we may say that even inanimate matter maintains itself in a continuous process similar to the growth of plants. Thus, recalling the ink-in-fluid model of the electron, we see that such a "particle" is to be understood as a recurrent stable order of unfoldment in which a certain form undergoing regular changes manifests again and again, but so rapidly that it appears to be in continuous existence. (Bohm 1980, p. 194)

A particle, in the implicate order framework, thus ought to be seen as "a recurrent stable order of unfoldment". The basic proposal is that biological organisms are basically similar:

> We may compare this to a forest, constituted of trees that are continually dying and being replaced by new ones. If it is considered on a long time-scale, this forest may be regarded likewise as a continuously existent but slowly changing entity. (Bohm 1980, p. 194)

Bohm's key suggestion then is that

> when understood through the implicate order, inanimate matter and living beings are seen to be, in certain key respects, basically similar in their modes of existence. (Bohm 1980)

We could ask here what is meant by "mode of existence". The atomistic idea is that particles exist continuously as particles. Bohm's idea is that such a continuously existing entity is just an abstraction we make from an underlying movement of unfoldment and enfoldment which is the more primary and fundamental reality. In this sense unfoldment–enfoldment is a more fundamental mode of existence, while continuous existence as a particle is something we, for convenience, attribute in situations where the order of unfoldment is recurrent and stable. The point is that there is a strong similarity between the more fundamental mode of being of inanimate matter and that of living matter. Both involve unfoldment and enfoldment. This opens up a new way to understand what happens when inanimate matter becomes animate matter. The key point is that the process of enfoldment and unfoldment that is taking place in inanimate matter is informed so that it starts producing a living plant:

> When inanimate matter is left to itself the above-described enfoldment and unfoldment just reproduces a similar form of inanimate matter, but when this is further "informed" by the seed, it begins to produce a living plant instead. Ultimately, this latter gives rise to a new seed, which allows the process to continue after the death of this plant. (Bohm 1980, p. 194)

Thus, life is defined as a particular kind of "well-informed" process of unfoldment and enfoldment. This underlines that there is no fundamental difference between inanimate matter and animate matter; the difference is rather in the nature of the enfoldment and unfoldment which is fundamental to both.

In other words, the underlying idea is that we have a basic process of enfoldment and unfoldment which gives rise to different structures, depending on how it is informed. Information thus plays an important *causal* role, which brings to mind Aristotelian ideas about the "formal" cause. As previously mentioned, this idea anticipates Bohm's later emphasis on the notion of *active information*, which we will discuss in more detail in Chap. 4.

What "inanimate matter" and "living matter" have in common is their "mode of existence" – for whether we talk about an inanimate entity (e.g. an electron) or a living entity (e.g. a tree), there has to be a certain recurrent stable order of unfoldment. The difference between an inanimate entity and a living entity then has to do with the *information* that acts on the order of unfoldment. In the case of living matter this is obviously more complex, but Bohm emphasizes the similarity:

> As the plant is formed, maintained and dissolved by the exchange of matter and energy with its environment, at which point can we say that there is a sharp distinction between what is alive and what is not? Clearly, a molecule of carbon dioxide that crosses a cell boundary into a leaf does not suddenly "come alive" nor does a molecule of oxygen suddenly "die" when it is released to the atmosphere. Rather, life itself has to be regarded

> as belonging in some sense to a totality, including plant and environment. (Bohm 1980, p. 194)

Here one might also say (in somewhat Bohmian jargon) that life has to do with a certain type of stable order of activity. Now, there is a doctrine called "functionalism" in the philosophy of mind which sees Aristotle as its historical predecessor. When considering the question of life, a functionalist would say that what is important is not so much the materials as such out of which a living organism is constituted, but rather the functional structure or organization of the organism. Thus the essence of life is the "form" or "functional structure". The organism has certain "structural properties", but what is important is the sorts of functions these structures perform. What makes a bunch of matter (with structural properties) a living organism is the fact that it is organized and functions in a particular way. I think that this is consonant with Bohm's approach, but unlike a typical functionalist he thinks that the origin of the organization is in the totality. He brings this out by comparing the growth of a plant with the electrons in the ink-in-dye analogy:

> It may indeed be said that life is enfolded in the totality and that, even when it is not manifest, it is somehow "implicit" in what we generally call a situation in which there is no life. We can illustrate this by considering the ensemble of all the atoms that are now in the environment but that are eventually going to constitute a plant that will grow from a certain seed. This ensemble is evidently, in certain key ways, similar to that ... of ink particles forming a droplet. In both cases the elements are bound together to contribute to a common end (in one case an ink droplet and in the other case a living plant). (Bohm 1980, p. 194)

One might think that Bohm implies that life can be reduced to inanimate matter, to the level where the laws of quantum theory prevail. He agrees that some features of life can be understood in this way (for example, the DNA molecule involves quantum theory, as Schrödinger famously anticipated; see Murphy and O'Neill (1995)). But Bohm wants to give life a much stronger status in his ontology:

> ... we are proposing that as the notion of the holomovement was enriched by going from three-dimensional to multidimensional implicate order and then to the vast "sea" of energy in "empty" space, so we may now enrich this notion further by saying that

> in its totality the holomovement includes the principle of life as well. (Bohm 1980, p. 195)

Of course, such an idea goes strongly against the prevalent views in contemporary biology, where one often meets the attitude that life can be reduced to fairly mechanical physical processes, such as natural selection at the macrolevel and the operation of the DNA molecule and the like at the microlevel. It seems that Bohm is attributing to "the principle of life" a similar status to that he attributes to the laws of nature. However, he is not arguing for vitalism, in the sense of claiming that life can be reduced to the operation of some non-physical force. Instead, "inanimate matter" and a non-physical "life force" are to be seen as features we can, if we wish, abstract from the holomovement, but which are, strictly speaking, not absolutely autonomous domains:

> Inanimate matter is then to be regarded as a relatively autonomous sub-totality in which, at least as far as we now know, life does not significantly manifest. That is to say, inanimate matter is a secondary, derivative, and particular abstraction from the holomovement (as would also be the notion of a "life force" entirely independent of matter). (Bohm 1980, p. 195)

Inanimate matter and life are thus not, strictly speaking, separate from each other (as they both have their ground in the holomovement and thus "meet" there). Nor does Bohm think that we can reduce life completely to the outcome of inanimate matter. Instead, as with everything else that exists, inanimate matter and life are the outcome of the holomovement, which is assumed to be primary, self-existent, and universal.

2.9 The Causal Architecture of the Holomovement

Having discussed the nature of matter and biological phenomena in terms of the implicate order, Bohm feels he is in a position to try characterize "the law of the holomovement" in more general terms. A basic feature of this law is that it is typically possible to abstract a set of relatively autonomous sub-totalities. The domain of the physical world in which the laws of classical physics apply is one example of such a sub-totality; inanimate matter more generally is another one; and life is yet another. Each such abstracted sub-totality then has laws that operate within it. Thus, there are, for example, Newton's and Maxwell's laws in the classical domain, quantum laws in the quantum domain, and certain biological regularities at the biological domain. Such laws are not absolute but operate under certain conditions and limitations that are defined only in a corresponding total situation. Bohm suggests that this operation generally has the following three features:

1. A set of implicate orders.
2. A special distinguished case of the above set, which constitutes an explicate order of manifestation.
3. A general relationship (or law) expressing a force of necessity which binds together a certain set of the elements of the implicate order in such a way that they contribute to a common explicate end (different from that to which another set of inter-penetrating and intermingling elements will contribute). (Bohm 1980, p. 195)

Let us first discuss what is meant by the "set of implicate orders". In the case of the ink-in-dye device, the set of implicate orders is the set of droplets in the situation in which the droplets have been enfolded one after the other with the cylinder rotated n times between each placing of the droplets. Each droplet then can be called an implicate order, and having many droplets means that there is a set of them. In the case of the electron, in terms of the implicate order, we think of the electron as a set of fields. Typically, each field is spread out and in this sense enfolded into the whole, thus constituting an implicate order. The fields differ in their degree of enfoldment; some may be spread out throughout space, some less so, and so on. In the case of the forest, the set of implicate orders is presumably all the seeds and all the matter in the environment that can potentially come together and constitute a plant, including the actual trees that constitute the forest at this moment.

Let us then consider the second feature, "the special distinguished case of the above set, which constitutes an explicate order of manifestation". In the case of the ink-in-dye analogy, the special case is, of course the droplet that is visible at this moment. In the case of the electron, it is that field out of the set of fields whose intensity is localized into a pulse, giving rise to a particle-like manifestation. In the case of the forest, it is the set of trees that make up the forest at this moment.

Finally, there is "the general relationship (or law) expressing a force of necessity which binds together a certain set of the elements of the implicate order in such a way that they contribute to a common explicate end (different from that to which another set of inter-penetrating and intermingling elements will contribute)". In the case of the ink-in-dye device, the overall necessity operates mechanically as the movement of fluid, according to the laws of hydrodynamics. In the case of the electron, the overall necessity is that of quantum laws, which operate without there being any mechanical medium "to do the causing". Finally, in the case of the forest, a key role is played by the information contained in the DNA molecule, which informs the molecules to make, say, an oak tree. Another kind of seed can then give rise to another kind of a tree.

We will see in the next chapter that Bohm extends a similar description even to psychological phenomena (for example, the way the elements of thought enfolded in the brain/mind come together to constitute a content of conscious experience at a given moment). He thus suggests that inanimate matter, life, and consciousness – all of which are thought to have their ground in the holomovement – have a similar *causal architecture*. This may open up a new way to think about the relationship of matter, life, and consciousness. These three are not assumed to be separate from each other, for they all have a common ground and thus "meet" at the holomovement in which they have their ground. And because the ground of each is, strictly speaking, in the holomovement, there is no need to reduce consciousness entirely to the brain, or the brain entirely to an arrangement of inanimate matter. Instead, a certain "democracy of sub-totalities" emerges. Inanimate matter, the brain, and consciousness are each, in the end, aspects that we have abstracted from the holomovement. Thus none of them enjoys the status of a substance into which everything else could be reduced. It is only the holomovement which is assumed to have such a status.

There is a tremendous tendency in contemporary science and even philosophy to favor matter over mind. This is perhaps understandable, because the explicate order of matter is the order that typically dominates our sensory experience. However, there are well-known troubles in the attempts to reduce consciousness entirely to the brain, and even the brain – or any living organism – entirely to inanimate matter. Bohm is suggesting that perhaps we have been taking inanimate matter too seriously, attributing to it too strong an ontological status. Physics itself is now undermining the view that inanimate matter ought to be seen as the ultimate substance. But classical physics was very successful in encouraging biologists and even psychologists to adopt such a reductionist idea. It is thus a major challenge for modern physicists to convince the biologists and psychologists and philosophers that they can stop believing in the absolute superiority of inanimate matter, and explore alternative concepts of reality. We have considered an alternative in which inanimate matter, life, and consciousness are all aspects from the underlying common ground in which they are united, and that there is no need to try to reduce them entirely into each other.

There is one more point about the nature of law in the holomovement that is worth mentioning again. This has to do with the origin of the force of necessity that prevails in a given situation. Bohm suggests that such an origin cannot be understood solely in terms of the explicate and implicate order that belong to the type of situation in question. For example, let us consider the origin of quantum laws. In a situation where quantum laws prevail, the implicate order is the wave function and the explicate order is the particle-like manifestation (or whatever it is that makes a spot appear in a photographic plate, or initiates the click of a Geiger counter). The wave function typically obeys the Schrödinger equation, which describes the main law that prevails in the situations. We cannot understand the origin of the Schrödinger equation by just analyzing the wave function (which is in an implicate order), or analyzing the spots in a photographic plate (which here reflect the explicate order, or the positions at which individual electrons have manifested themselves as particles). We need to go into a broader and deeper context, if we are to try to understand why the Schrödinger equation (rather than some other equation) applies at the quantum domain.

Bohm suggests that we have to accept a force of necessity as *inherent* in the overall situation under discussion. Thus, the Schrödinger equation has to be accepted as inherent in the quantum domain. He thinks that for any force of necessity it is the case that

> [a]n understanding of its origin would take us to a deeper, more comprehensive and more inward level of relative autonomy which, however, would also have its implicate and explicate orders and a correspondingly deeper and more inward force of necessity that would bring about their transformation into each other. (Bohm 1980, pp. 195–196)

We can again see here how Bohm thinks that there is no need to assume that there exists a fundamental level – an idea which he proposed in his 1957 book *Causality and Chance in Modern Physics*. There may well be a level of nature that lies deeper than the quantum level, but Bohm sees no reason to assume that this would be the fundamental level. This suggests that *the laws of physics should always be seen as relative to a particular context, in which they ought to be seen as inherent features, not explainable within the context.* However, as no ultimate level or context is assumed, no law of nature is assumed to be absolutely inherent. For any law, it is in principle assumed to be possible to understand its origin, should one be able to go beyond the context in which it applies.

Of course, the above raises the question how we are to think of a law that might prevail in the "totality of all that exists". For this totality contains, by definition, everything, and thus it is not possible to go outside of it. Thus, any law that prevails in it would be absolutely inherent. However, to put it mildly, it is not likely that we will ever know what the "totality of all that exists" actually contains, and thus the question about the law that might prevail in it is a mere theoretical curiosity.

We have now examined in some detail how the implicate order applies in matter, and how it may be extended to biological phenomena. In the next chapter we will examine whether the implicate order can be extended to psychological phenomena. We will then consider whether it is possible to understand in a new way the relationship between matter and consciousness in the context of the implicate order framework.

3

The Architecture of Consciousness

3.1 Introduction

Matter and consciousness have traditionally been seen as two entirely different substances, most notably by Descartes, who defined matter as extended substance (*res extensa*) and consciousness as thinking substance (*res cogitans*).[1] In the philosophical tradition, to call something a substance is to give it a very fundamental status indeed. A substance is something that can exist even if nothing else exists. In this sense a substance is a bit like a world in itself. Thus to say that matter and consciousness are two different substances is a bit like saying that there are two different worlds which can keep on existing even if the other world did not exist. Such independence of the existence of something of the existence of another thing is called in philosophy "ontological independence".[2]

[1] I am using the word "consciousness" for Descartes' "thinking substance". According to modern interpreters, Descartes' notion of thinking substance is broad, and corresponds fairly well to the way the word "consciousness" is used in the contemporary discussion. It includes not just abstract conceptual thought but also desires, feeling, will, etc.

[2] The word "ontology" comes from the Greek *logos peri ta onta* where *ta onta* means "the beings", *logos* has meanings like "word", "speech", "reason", and *peri* means "around", and perhaps in this context "about". So ontology originally would have meant something like "reasoning about the beings" (S. Vesterinen, private communication). In contemporary philosophy "ontology" likewise refers to the study of being. It is often thought to be equivalent to "general metaphysics", which studies the general and basic features of being, as opposed to "special metaphysics", which traditionally studies more specific topics such as "nature", "soul", and "God". The German 17th-century "school-metaphysician" Göckel was the first philosopher to call general metaphysics ontology, a usage established in the 18th century by Wolff (Juti 2001).

To say that two things are ontologically independent does not, however, need to imply that they cannot affect each other. Thus they could depend upon each other *causally*, so that, for example, changes in one substance could bring about changes in the other substance, and vice versa. So it seems that ontological independence of two substances is consistent with their causal dependence (cf. Niiniluoto (1987)).

Indeed, it seems obvious – and was to Descartes – that our consciousness and the matter constituting our bodies are not causally independent of each other but are instead very closely related, and seem to affect each other. Descartes described the relation of consciousness and body as a mutual *interaction* of the two substances. But because the substances were assumed to be so different, it was very difficult to understand *how* they could interact. Descartes thought that they have nothing in common, while the understanding of their relationship would seem to require that they have at least one thing in common, namely the link through which they interact. As no such link was available in the assumed properties of matter and consciousness, Descartes assumed that it was God, the ultimate substance and the creator of the two other substances matter and consciousness, who was responsible for their interaction (Cottingham 1986).

As science and philosophy developed, more and more people felt that such an appeal to God was not really acceptable, and thus gave it up. However, with the disappearance of such appeal there also disappeared the explanation of how the two very different substances could interact. This problem about the nature of matter and consciousness and their relationship is known in philosophy as the *mind–body problem* (sometimes also called the ontological problem (Churchland 1988), as it concerns ontological questions such as "what is consciousness", "what is matter", "what is the nature of their relation", etc.). While some philosophers have from time to time suggested that the problem is a reflection of, say, a sloppy use of language rather than a genuine problem, many others think the problem is a genuine one and needs and deserves a solution.

In the course of history many solutions have been proposed. For example, Spinoza pointed out that because Descartes assumed that God was the primary substance, he was, strictly speaking, mistaken to call matter and consciousness substances. Thus Spinoza proposed that there is only one substance and that matter and consciousness are two different aspects of this substance, those that are known to human beings. Leibniz, in turn, proposed that reality consists of monads, which are not material entities but rather mental, perceiving entities. Matter has the status of appearance, while consciousness is what is real. Leibniz thus favored consciousness to matter, and his viewpoint can be seen as a form of phenomenalism or idealism (although it is often, and perhaps mistakenly, thought to be a parallellist form of dualism; see McCann (1994, pp. 341–342)).

One option, adopted by both Spinoza and Leibniz, for example, is thus to give up the idea that there are *two different* fundamental substances. Instead, one could assume that there is only one substance that is the fundamental "stuff" of the universe, and that everything else that exists can in some way be derived from this fundamental stuff.[3] In recent times it has been common to give up the idea that consciousness is a substance that can exist independently. Thus modern science and philosophy have often opted for the idea that matter is the fundamental stuff of the universe, while consciousness is something that does not exist or else can be derived from matter. Matter in this context has usually been assumed to be something fairly mechanical, for example something that in the end consists of some basic building blocks or "elementary particles", which combine and interact mechanically to produce higher-level structures and activities.

There are a number of different materialist viewpoints. These have been extensively discussed in recent literature, so I will only describe them very briefly here, to help those readers who are not familiar with the literature (for a more thorough description see, for instance, Churchland (1988), Searle (1992), or Seager (1991)).

There is *emergent materialism*, which assumes that consciousness is a property that has emerged from matter at some point in biological evolution, but once emerged, can no longer be reduced to matter. In so far as emergent materialism assumes that consciousness cannot be reduced to matter, it is also often called property dualism, as it has a basically dualistic flavor.

There is *behaviorism*, which says that talk of consciousness can be reduced to talk of actual or possible behavior, and that there is no need to assume that consciousness exists in any autonomous sense. Behaviorism was an attempt to study the mind by studying behavior in a very systematic sense, focusing on the relations between stimulus and response. However, the focus on external behavior, though valuable, is

[3] It is often thought that Leibniz was a pluralist, someone who believes that there exist a great number of substances, as he thought that each monad is a substance in its own right. However, the key point is that, strictly speaking, he postulated only one type of substance, the monad, and in this sense his ontology is monistic, rather than dualistic or pluralistic. See McCann (1994).

today widely thought to have been much too one-sided, as it ignores our inner conscious experiences.

There is the *mind–brain identity theory*, which says that consciousness is identical with certain neurophysiological processes. One is no longer denying the existence of consciousness (as stronger versions of behaviorism did); one is instead saying that consciousness is real, but that it is identical with physical states. The stronger version is *type identity theory*, which assumes that types of mental states (e.g. pain) are identical to types of neural activity (e.g. C-fiber firing). The weaker version is *token identity theory*, which says that a given mental state is identical with some physical state but is more liberal as to what that physical state might be (leaving it open, for example, that extraterrestrials with a physical constitution different from ours could have the same type of pain as we do, an option excluded by the type identity theory).

Closely associated with token identity theory is *functionalism*. As just mentioned, the former allows that systems which have a different physical constitution could all "instantiate" or "realize" the same type of mental state (e.g. pain). But, one might think, surely those different systems must have something in common which enables them to realize the same state. Functionalism suggests that regardless of their different physical constitution, these systems perform the same function. When I am in pain, there are a number of physiological activities taking place in my body and brain. A certain input (for instance, a flame of a candle) activates certain nerve cells, which in turn activate other nerve cells, leading to the activation of motor neurons (e.g. I withdraw my hand). Functionalism in its simplest form assumes that as long as in some system (regardless of its specific material constitution) certain functions are performed that are equivalent to those that are performed in me when I have pain, then that system also has pain.

Functionalism thus assumes that consciousness is identical with a set of certain more abstract "functional roles" or "functional structures", which can, at least in principle, be realized in different physical systems (e.g. human beings, sufficiently complex robots, perhaps extraterrestrials that differ from us physico-chemically but are functionally equivalent). A familiar example of such a functional structure is a computer program. The same program (software) can be implemented in different physical computers (hardware), a feature that is known as "multiple realization". Analogously, functionalism assumes that mental states are "multiply realizable".

Yet another doctrine is *instrumentalism*, which says that although it can be useful to refer to consciousness when explaining behavior, strictly speaking all that exists are purely physical processes. Instrumentalism is a sophisticated doctrine, but it seems to have a strange bias in favour of the physical. Why not adopt the instrumentalist attitude "all the way down"? Why give the physical sciences such a strong status?

Finally, there is *eliminative materialism*, which holds that some of our usual ideas about consciousness are simply mistaken and that instead of trying to relate such ideas to matter, we should simply give them up and proceed with empirical research in cognitive neuroscience. Like instrumentalism, eliminative materialism is also a very sophisticated doctrine. It has a very progressive flavour as it advocates the view that we ought to be prepared to give up dearly held views, if scientific evidence calls for this. It radically questions our traditional views about consciousness, which it thinks constitute a theory of "folk psychology", a theory that is false.

I think, however, that there are at least two problems with contemporary eliminative materialism. Firstly, there may well be more truth to our common-sense intuitions about conscious experience than an eliminative materialist is prepared to admit. Secondly, when it comes to the issue of what should be the new theory about the physical nature of consciousness, it seems that eliminative materialists are not radical enough. Modern physics strongly suggests that the more fundamental physical processes are holistic. Brains, too, have the ground of their existence in these holistic physical processes; as a result, a scientific approach to the brain cannot just ignore the possible role of these more fundamental physical processes. Thus, a truly radical "eliminative materialist" would seriously consider the nature of the physical context in which brains actually exist, and when doing this, he or she might well be led to give up some dearly held mechanistic views about the nature of the neurophysiological processes, or at least try to see their limits of applicability.

In the contemporary debates, however, eliminativists have typically adopted a fairly conservative attitude when it comes to the question of the physical nature of consciousness (see, for example, Grush and Churchland's (1995) debate with Penrose and Hameroff (1995); see also Hameroff and Penrose (1996) and Hagan et al. (2002)). Thus although eliminative materialism may in some ways be a radical view about the nature of consciousness, it can in its present form also be seen as a conservative form of materialism which is trying to protect standard mechanistic neuroscience from being questioned or "eliminated" by the new developments in physics. What we need is an "eliminativism" that "goes all the way" with science and is prepared to give up all dearly held views (and not just views about conscious experience), if scientific evidence suggests the need to do this. Also, elimination is not the only response to a concept or theory that does not work in a certain domain. As we have emphasized, concepts tend to have a limited range of applicability. Thus in many cases we do not so much need to eliminate concepts and theories altogether, but rather to see their limits of applicability.

Let us move on. One way of approaching the question of the relationship between matter and consciousness is to focus upon the type of relation that is assumed to hold between them. We have already mentioned *reduction*, which typically suggests that consciousness can be reduced to some physical processes (e.g. neural processes), leading to the postulation of some kind of *identity* between consciousness and some parts of matter. We also mentioned *emergence*, which is typically used in order to say that consciousness has emerged from matter. Some philosophers, like Searle, have emphasized *causation*, suggesting that consciousness is a causal, higher-level effect of certain processes in the brain. Finally, much discussion has recently centered around a relation called *supervenience*, which usually means that although consciousness cannot be reduced to physical processes, it is strongly dependent upon them. Often this dependence is described by saying that there cannot be a change in consciousness (the supervenient property) without there being a correlated change in the underlying physical substrate (the "subvenient" base) (see Kim (1999)).

Within these viewpoints and relations there are variations, and there are other materialist viewpoints not mentioned above. While these views no doubt provide us with valuable insights into the nature of consciousness and its relationship to matter, many philosophers feel that they in an important sense leave consciousness unexplained. Emergent materialism typically does not explain how consciousness arises from matter, and in its strongest version even assumes that such explanation is impossible. It is thus fairly close to Cartesian substance dualism and inherits its problems. Behaviorism is often thought just to ignore the obvious facts about consciousness, and is thus no longer considered a plausible explanation of consciousness. The mind–brain identity theory (both the type and token versions) does not explain how consciousness and some neurophysiological processes can be the same thing, and thus the claim that they are "identical" is not really an illuminating explanation. Functionalism typically allows that consciousness could be realized, in principle, by a set of mechanical functions carried out by mechanical components and structures. The trouble with this is that it is left as a mystery *why* performing any set of such mechanical functions should give rise to consciousness. Instrumentalism is often felt to give too weak a status to consciousness, while many feel that eliminative materialism goes too far in claiming that our views about consciousness are mistaken and ought to be eliminated. As Thomas Nagel has succinctly put it, it seems that neither dualism nor materialism can be correct solutions of the mind–body problem, but at the same time it is not clear what the alternative could be (Nagel 1995). For a very illuminating discussion of these viewpoints and their problems, see also Seager (1991).

3.2 Consciousness and the Implicate Order

With the above in mind as a background, let us now go on to consider how Bohm tries to extend the notion of implicate order to consciousness. His proposal is simply that consciousness (which he takes to include thought, feeling, desire, will, etc.) is to be comprehended in terms of the implicate order, along with reality as a whole. He thus suggests that

> the implicate order applies both to matter (living and non-living) and to consciousness, and that it can therefore make possible an understanding of the general relationship of these two, from which we may be able to come to some notion of a common ground of both (rather as was also suggested ... of the relationship of inanimate matter and life). (Bohm 1980, p. 196)

The idea here is, firstly, that matter and consciousness have, fundamentally, the same order or architecture. It seems that Bohm thinks that this, in turn, makes it possible and reasonable to develop a notion of a "common ground" of matter and consciousness. This idea of a "common ground" suggests that, from a philosophical point of view, he advocates a view called "neutral monism", which resembles Spinoza's view mentioned above. The idea in neutral monism is that the fundamental reality is neither mind (idealism) nor matter (materialism) nor both (dualism) but rather a "neutral" ground beyond mind and matter. In contrast, as we saw above, it seems that the majority of researchers in philosophy of mind are, in one way or another, trying to incorporate mind into a materialist metaphysics. It is of course interesting to note here that Bohm who is a physicist (and thus might be thought to have a bias in favor of materialist explanations) is not trying to do this.

He is, however, aware of the difficulties of relating mind and matter, and thus at least implicitly admits that it is not obvious how the implicate order could help us to understand their relationship:

> To obtain an understanding of the relationship of matter and consciousness has... thus far proved to be extremely difficult, and this difficulty has its root in the very great difference in their basic qualities, as they present themselves in our experience. (Bohm 1980, p. 196)

Bohm thinks that Descartes expressed the difference with particularly great clarity, with his description of matter as "extended substance" and consciousness as "thinking substance". He sees a clear connection between Descartes' "extended substance" and his own notion of explicate order:

> Evidently, by "extended substance" Descartes meant something made up of distinct forms existing in space, in an order of extension and separations basically similar to the one that we have been calling explicate. (Bohm 1980, p. 196)

It seems fair to say that much of contemporary philosophy of mind, cognitive science, etc. is in some sense trying to reduce consciousness to "extended substance" or the "explicate order". Theories that identify types of conscious experience (e.g. pain) to types of neural activity (e.g. C-fibers firing) obviously try to do this. But even functionalism tries to do this, for functionalists typically assume that the functional structure or "computer program" that is thought to be the essence of a certain conscious experience (e.g. pain) could in principle be implemented to (and in a sense be identical with) a neurophysiological state.[4] In contrast to this, it is interesting to note that Descartes and Bohm agree in that conscious experience cannot be reduced to the explicate order. Bohm writes:

> By using the term "thinking substance" in such sharp contrast to "extended substance", [Descartes] was clearly implying that the various distinct forms appearing in thought do not have

[4] Note, however, that there might be an interesting relationship between functionalism and the implicate order. Strictly speaking, functionalism identifies types of mental states with types of functional architecture; so it thinks that what is essential about mental states is order and structure, rather than the existence of objects in the explicate order. See, for example, Lycan (2003).

> their existence in such an order of extension and separation (i.e., some kind of space), but rather in a different order, in which extension and separation have no fundamental significance. The implicate order has just this latter quality, so in a certain sense Descartes was perhaps anticipating that consciousness has to be understood in terms of an order that is closer to the implicate than it is to the explicate. (Bohm 1980, pp. 196–7)

This is, of course, a very interesting possibility. Bohm is suggesting that contemporary philosophers of mind have chosen the wrong track in so far as they are trying to explain consciousness in terms of the explicate order. At the same time, he is suggesting that when it comes to consciousness, Descartes was on the right track. This is very interesting because Descartes' views on consciousness are often dismissed. Almost every textbook in contemporary philosophy of mind begins with a brief discussion of Descartes' dualism, where his views are dismissed. In contrast, Bohm suggests that Descartes realized something very important about the essence of consciousness, namely that our thoughts live in an order in which extension and separation are not fundamental. However, Bohm is also critical of Descartes' views. For he notes that when we start, as did Descartes, with extension and separation as primary for matter, then we can see nothing in this notion that can serve as a basis for a relationship between matter and consciousness, whose orders are so different:

> Descartes clearly understood this difficulty and indeed proposed to resolve it by means of the idea that such a relationship is made possible by God, who being outside of and beyond matter and consciousness (both of which He has indeed created) is able to give the latter "clear and distinct notions" that are currently applicable to the former. Since then, the idea that God takes care of this requirement has generally been abandoned, but it has not commonly been noticed that thereby the possibility of comprehending the relationship between matter and consciousness has collapsed. (Bohm 1980, p. 197)

Bohm now reminds us that he has shown in some detail that matter as a whole can be understood in terms of the notion that the implicate order is the immediate and primary actuality (while the explicate order can be derived as a particular, distinguished case of the implicate order).

> The question that arises here, then, is that of whether or not (as was in a certain sense anticipated by Descartes) the actual

> "substance" of consciousness can be understood in terms of the notion that the implicate order is also its primary and immediate actuality. (Bohm 1980, p. 197)

It is clear what Bohm is driving at. Descartes thought that consciousness involved something like the implicate order, while matter was assumed to essentially involve the explicate order. Bohm suggests that Descartes was on the right track about consciousness, but on the wrong track when proposing that the explicate order is fundamental to matter. Quantum theory and relativity suggest that the implicate order is fundamental for matter, thus implying that consciousness and matter have in common the same order or architecture. This, in turn, might make it possible to understand their relationship. Bohm thinks that the way to do this is to assume that consciousness and matter, though different aspects of reality, nevertheless have a common ground in which they "meet":

> If matter and consciousness could in this way be understood together, in terms of the same general notion of order, the way would be opened to comprehending their relationship on the basis of some common ground. Thus we could come to the germ of a new notion of unbroken wholeness, in which consciousness is no longer to be fundamentally separated from matter. (Bohm 1980, p. 197)

It seems that the first question for Bohm is whether we can understand matter and consciousness in terms of the same notion of order. This, he implies, would then make it reasonable to seek some common ground for both.

It is interesting to contrast Bohm's view with those of contemporary physicalists. It seems that the majority of physicalists are still assuming that the physical world is roughly the same as "*res extensa*" or "explicate order", and quite a few assume that consciousness should be reduced to the explicate order or else shown to be emergent from it, and/or causally produced by it, and/or strongly supervenient upon it. Bohm disagrees on all points. First of all, he argues that although the physical world has a domain in which the explicate order applies, the more fundamental nature of matter cannot be understood in terms of the explicate order, but we need instead the implicate order. Further, he suggests that the more fundamental nature of the mind and conscious experience likewise has to be understood in terms of the implicate order, although mind (just like matter) also has a domain which can be understood in terms of the explicate order. Just like the physicalists,

Bohm seeks to integrate mind and conscious experience within the view of the world given by the natural sciences. But there is a striking difference in the way this is attempted. The mainstream assumes that the order of matter is basically mechanistic and then tries to integrate mind and conscious experience within the mechanistic picture, with well-known difficulties. In contrast, Bohm is suggesting that the order of matter is fundamentally holistic; and this opens up the possibility of integrating mind and conscious experience to our scientific picture of the world, without denying their well-known holistic features.

What Bohm and the physicalists thus have in common is a certain search for ontological unity. Yet they differ radically in the way in which they attempt to achieve this unity. Bohm's approach takes physics, the most fundamental of the natural sciences, as his starting point, and tries to articulate a general view which does justice to the results of relativity and quantum physics. It is then in the framework of this view that he discusses the relationship between mind and matter. In contrast, it seems that many contemporary physicalists often leave out what modern physics has to say about the physical world; they rely on older, more familiar mechanistic views about the physical world and proceed to discuss the relationship between mind and matter in the context of such a view.

It is also interesting that Bohm views Descartes rather differently from contemporary philosophy, as indeed already mentioned. Contemporary physicalist philosophy of mind often assumes that Descartes was correct (or at least on the right track) in his view about matter, but mistaken in his view about the mind. Bohm turns this upside down. For he suggests firstly that Descartes was mistaken in assuming that the essence of matter is extension, and thus the explicate order (although Bohm admits that Descartes' view of matter, in some respects, fits approximately with the domain of classical physics). Secondly, Bohm suggests that Descartes was correct or on the right track about the mind; correct in emphasizing that the essence of the mind cannot be understood in terms of extension, or the explicate order, and correct in implying the need for some other notion of order. One might here note that Bohm thinks that there is indeed a part of the mind that can be understood in terms of the explicate order (e.g. our usual sensory experience of a "virtual reality" in three-dimensional space). Yet, as we shall see later in this chapter, he emphasizes that it is a mistake to think that this explicate order that we experience in consciousness is the essence of consciousness. Rather, it ought to be seen as a relatively autonomous sub-totality that arises from the underlying implicate order of conscious experience, just as the explicate order of classical physics is a relatively autonomous sub-totality that arises from the implicate order of matter. Bohm thus emphasizes that the deeper essence of conscious experience has to be understood in terms of the implicate order, and in this respect he is perhaps in some ways in agreement with Descartes.

Figure 3.1 illustrates very schematically how Descartes, contemporary physicalism, and Bohm, respectively, think of matter and consciousness in relation to the explicate and implicate orders.

Note, however, that functionalism is closer to Bohm's view than, say, the identity theory. For functionalism identifies mental state types (e.g. pain) with abstract functional structures. Most functionalists accept so-called "token identity", which means that they assume that any particular pain is identical with a particular physical state of an organism. The abstract functional structure is thought to be "implemented" or "realized" in the physical system. In this sense, instances of pain "live" in the explicate order (i.e. in the physical brain state). However, pain as a type is something more abstract, and this resembles somewhat the idea that the implicate order is the essence of consciousness.

3.3 Does the Implicate Order Prevail in Conscious Experience?

How then does Bohm justify, in more detail, his idea that the implicate order applies in consciousness? He starts by reminding us of the fact that

> various energies such as light, sound, etc., are continually enfolding information in principle concerning the entire universe of matter into each region of space. (Bohm 1980, p. 197).

This is one of the essential features of the implicate as opposed to the explicate order of matter. How is this relevant to consciousness?

> ...such information may of course enter our sense organs, to go on through the nervous system to the brain. More deeply all the matter in our bodies, from the very first, enfolds the universe in some way. Is this enfolded structure, both of information and of matter (e.g., in the brain and nervous system), that which primarily enters consciousness? (Bohm 1980, pp. 197–8)

There are two suggestions here. One is fairly conventional and obvious. If I want to perceive the furniture in the room I am in, I do not need to transport the tables and chairs into my brain; it is enough that light

Descartes

	Explicate order	Implicate order
Matter	X	
Consciousness		X

Contemporary physicalism

	Explicate order	Implicate order
Matter	X	
Consciousness	X	

Bohm

	Explicate order	Implicate order
Matter	x	X
Consciousness	x	X

x = derivative part; X = fundamental part

Fig. 3.1. Matter and consciousness in relation to the explicate and implicate orders for Descartes, contemporary physicalism, and Bohm which at each small region of the room "enfolds information" about the order of the whole room enters the pupil of the eye, and so on. The second suggestion is much more radical and exotic. The material particles that constitute my body and brain can be looked at quantum field theoretically; and when this is done in a Bohmian fashion, one can say that these constituents enfold the universe in some way (which presumably means that they contain information about the whole universe, and that information about each one of them can be found everywhere in the universe – just as in a hologram each part of the plate contains information about the whole image, and information about each part of the image can be found everywhere). What would it mean for that latter "enfolded structure" to enter consciousness? Would it give rise to a sense of the existence of the whole universe in some way? This opens up a fascinating possibility of "Bohmian existentialism", which we will consider later in this chapter. For now, let us discuss whether information is actually enfolded in brain cells. Bohm refers to Pribram's (1971) work on brain structure:

> Pribram has given evidence backing up his suggestion that memories are generally recorded all over the brain, in such a way that information concerning a given object or quality is not stored in a particular cell or localized part of the brain but rather that all the information is enfolded over the whole. This storage resembles a hologram in its function, but its actual structure is much more complex. (Bohm 1980, p. 198)

The above text by Bohm was published in 1980, and we could ask how well Pribram's work stands up today. My discussions with some cognitive neuroscientists suggest that it still deserves to be explored in more detail. Also, there has been intensive work in artificial neural networks since 1980, and some of this work resembles and might support the holographic theory of memory (see, for instance, Kohonen 1984). However, let us proceed to consider the implications the holographic view of memory, if correct, would have:

> We can then suggest that when the "holographic" record in the brain is suitably activated, the response is to create a pattern of nervous energy constituting a partial experience similar to that which produced the "hologram" in the first place. (Bohm 1980, p. 198)

This is of course similar to the way holograms usually work. The hologram, as it were, "freezes" the light front, and when light is shone on it again, this light front is recreated. The reason we see a three-dimensional image is that the hologram (unlike an ordinary photograph) records information about the phase. The idea here is that the brain stores in a holographic fashion information about the experience we are having at a particular moment, and this experience can be recreated by activating the record. Bohm next elaborates this idea:

> But [the partial experience] is also different [from the original experience] in that it is less detailed, in that memories from many different times may merge together, and in that memories may be connected by association and by logical thought to give a certain further order to the whole pattern. In addition, if sensory data is also being attended to at the same time, the whole of this response from memory will, in general, fuse with the nervous excitation coming from the senses to give rise to an overall experience in which memory, logic, and sensory activity combine into a single unanalysable whole. (Bohm 1980, p. 198)

This provides a nice way of thinking about the old philosophical battle between "rationalism" and "empiricism", and suggests a middle way not unlike that advocated by Kant (although there are, of course, important differences between Kant's and Bohm's views). The idea here is that our overall experience is typically a *fusion* of sensory activity, memory, and logic. What makes the fusion possible is the fact that the information that is the ground of the experience exists as patterns of nervous energy, whether this is information in the memory or incoming sensory information. The key point is that, in principle, it is possible that such patterns can be added together, thus making intelligible the idea that our overall experience could involve such a fusion. One can here think of the interference of waves, such as water waves. A given wave pattern can be a fusion of many different waves in a region. Bohm emphasizes here that the overall experience is a "single unanalysable whole". This indeed is the way it feels to the "subject" – it is not easy when considering one's overall experience to separate the contributions made by the senses, memory, and logic from each other.[5]

The above description is somewhat Kantian in spirit, as it suggests that the information that is already in the brain can play a key role in shaping our experience. Thus, in principle, the brain could supply

[5] Of course, when dealing with wave patterns in physics, one can use Fourier analysis to break wave patterns into their components and Fourier synthesis to construct complex wave patterns out of simple waves. This perhaps suggests a research program of "Fourier analysis" for cognitive science, if one thinks that such analysis of experience to its components would be relevant.

the "forms of perception" (space and time) and the "categories of understanding" (e.g. causality), which could then be used to shape sensory input and make it into the "virtual reality" of a spatio-temporal conscious experience in which certain regularities (e.g. causality) prevail. Kant, however, did not think that the forms of perception and the categories of understanding are neural processes. What the above suggests is thus – very preliminarily – a way to "naturalize" parts of Kantian philosophy (cf. Petitot et al.' s (1999) project of "naturalizing phenomenology").

Bohm was more open than Kant to studying the role of the brain in constructing our experience. Given the 20th-century advances in neuroscience, he was, of course, in a different position to consider this role. However, like Kant, Bohm, too, thought that the essence of conscious experience cannot be entirely reduced to mechanical processes in the brain, as is brought out by the following passage:

> Of course, consciousness is more than what has been described above. It also involves awareness, attention, perception, acts of understanding, and perhaps yet more. We have suggested . . . that these must go beyond a mechanistic response (such as that which the holographic model of brain function would by itself imply). So in studying them we may be coming closer to the essence of actual conscious experience than is possible merely by discussing patterns of excitation of the sensory nerves and how they may be recorded in memory.
>
> It is difficult to say much about faculties as subtle as these. However, by reflecting on and giving careful attention to what happens in certain experiences, one can obtain valuable clues. (Bohm 1980, p. 198)

Here Bohm reveals his intuitions about the nature of consciousness. He proposes that conscious experience involves "non-mechanistic" aspects. In effect, he advocates a kind of "phenomenological method" to study these aspects. We have to "reflect on" and "give careful attention" to what happens in certain experiences. This resonates with the views not only of the phenomenologists, but also of contemporary consciousness researchers like Antti Revonsuo, who although emphasizing the role of neuroscience, also emphasizes the need to take our inner experiences seriously in consciousness studies. In contemporary literature this is often described by saying that we need to adopt the "first-person perspective" when studying consciousness (in addition to the "third-person perspective", which we typically use in the natural sciences). Bohm in the previous passage is clearly advocating the use of the first-person perspective in the study of consciousness, while also underlining the importance of neural studies which can be done from the third-person perspective. Such combination of first- and third-person perspectives is, of course, one of the key characteristics of the new academic field of consciousness studies. Notice that the text above was published in 1980, somewhat before the recent resurgence of consciousness studies (e.g. Baars (1988), Dennett (1991), Searle (1992), Penrose (1989, 1994), Crick (1994), Lycan (1987), Chalmers (1996)). Thus it seems that Bohm was directly addressing conscious experience in his research, while most analytical philosophers, psychologists, cognitive neuroscientists, cognitive scientists, etc. were still focusing on other aspects of the mind. Of course, conscious experience has been systematically and continuously studied in phenomenology and in the various developments in "continental philosophy" which build upon phenomenology. As we will see later, there are some interesting similarities between, say, the views of Husserl and Bohm about time consciousness, for example.

Bohm's favourite example when considering the more subtle aspects of consciousness is our experience when listening to music. This example will be important to us later on in Chap. 5, when we are comparing Bohm's views with some of the views in the current debate on time consciousness. Let us thus examine it here in some detail.

> Consider, for example, what takes place when one is listening to music. At a given moment a certain note is being played but a number of the previous notes are still "reverberating" in consciousness. Close attention will show that it is the simultaneous presence and activity of all these reverberations that is responsible for the direct and immediately felt sense of movement, flow, and continuity. To hear a set of notes so far apart in time that there is no such reverberation will destroy altogether the sense of a whole unbroken, living movement that gives meaning and force to what is heard. (Bohm 1980, pp. 198–199)

A number of different aspects of musical experience come out in the above passage. First of all, musical experience has a temporal structure, in the sense that we can distinguish between "present", "past", and "future". The "present" involves the note that is being heard for the first time "now". The "past" includes those notes that were first heard some time ago, and still reverberate. The "future" involves the expectations of what will be heard next. Secondly, Bohm suggests that musical experience is characteristic of a "direct and immediately felt sense of movement, flow, and continuity". He explains this by referring to the simultaneous presence and activity of the reverberations. There is thus a sense in which the "past" is in the "present" and is in a sense perceived – a feature of conscious experience which also puzzled the famous phenomenologist Edmund Husserl. Finally, Bohm notes that the time interval between the notes makes a difference to musical experience. The sense of movement requires that the interval is not too large. In other words, there is a certain "temporal window" within which the notes have to be heard if there is to be a sense of movement.

But what are these "reverberations" and how do we experience them? Are they memories, re-presentations of the notes heard or presented earlier? According to Bohm, no:

> ... one does not experience the actuality of this whole movement by "holding on" to the past, with the aid of a memory of the sequence of notes, and comparing this past with the present. (Bohm 1980, p. 199)

This is in line with Husserlian phenomenology, and indeed seems reasonable. There seems to be a sense in which a reverberation is not a memory. But if it is not a memory, then what is it? Bohm answers:

> ... the "reverberations" that make such an experience possible are not memories but are rather active transformations of what came earlier. (Bohm 1980, p. 199)

So the idea is that those notes which have been presented in conscious experience continue to live their life as present (rather than represented) elements while being actively transformed. What happens in such a transformation? First of all, there is

> a generally diffused sense of the original sounds, with an intensity that falls off, according to the time elapsed since they were picked up by the ear. (Bohm 1980, p. 199)

This seems fair enough and accounts for the experienced difference between the notes played at a given moment and the reverberations of previous notes. After all, there is a difference in our experience when the C major chord tones C-E-G are all played simultaneously and when they are played one after the other. The active transformation where the intensity of the sound falls off as a function of time is responsible for the fact that we do not just hear a chord in the latter case by the time G is played. Rather, we hear G while C and E are simultaneously reverberating as active transformations with a lower intensity. Of course, this is also an experience of a C major chord but not the same as the one we have when the tones are played simultaneously.

Bohm further emphasizes that such active transformations involve

> various emotional responses, bodily sensations, incipient muscular movements, and the evocation of a wide range of yet further meanings, often of great subtlety. (Bohm 1980, p. 199)

This is an important reminder that "time consciousness" (and conscious experience in general) includes not just cognition in the narrow sense but also emotion and bodily sensations.

When we are listening to music, we are typically taking in a sequence of notes. Each note undergoes an active transformation that involves a sense of the sound and, together with other sounds, gives rise to the emotional, bodily, and muscular responses. Many levels of consciousness are thus involved. These transformations interpenetrate and intermingle, and it is such co-presence, according to Bohm, which gives rise to "an immediate and primary feeling of movement" essential to music.

Our main question in this chapter is whether the implicate order prevails in conscious experience, and we are now in position to consider whether it prevails in a typical musical experience. It indeed seems to do so. Consider the ink-in-fluid model of an electron, in which the latter is conceived as

> a co-present set of differently transformed ensembles that interpenetrate and intermingle in their various degrees of enfoldment. (Bohm 1980, p. 199)

Analogously, in musical experience there is also a co-present set of differently transformed ensembles that interpenetrate and intermingle in their various degrees of enfoldment. Bohm's conclusion is that

> [t]his activity of consciousness evidently constitutes a striking parallel to the activity that we have proposed to the implicate order in general. (Bohm 1980, p. 199)

In other words, what an electron and a musical experience have in common is that both essentially involve *co-present elements in different degrees of enfoldment* – which is an important characteristic of the implicate order, or one of the key "marks" of the implicate order.

Bohm elaborates on the analogy:

> In such enfoldment [of the droplets], there is radical change, not only of form but also of structure, in the entire set of ensembles (which change we have... called a metamorphosis); and yet, a certain totality of order in the ensembles remains invariant, in

> the sense that in all these changes a subtle but fundamental similarity of order is preserved. (E.g. a linearly ordered array of droplets may be enfolded together in such a way that this order is still subtly held in the whole set of ensembles of ink particles.) In the music, there is, as we have seen, a basically similar transformation (of notes) in which a certain order can also be seen to be preserved. (Bohm 1980, p. 199)

Enfoldment thus does not completely destroy an order, it just transforms it into a different form. An explicate order is transformed into an implicate order via enfoldment; unfoldment does the reverse. Bohm suggests that such a preservation of order takes place even in the fundamental processes in the domain of inanimate matter. Indeed, he assumes that such processes are essentially processes of enfoldment and unfoldment, and it is this that gives rise to the effects that we observe, and also explains some puzzling features (such as the discontinuity of motion). The holomovement is thus able to preserve a tremendous amount of order within it. Likewise, in musical experience a certain order is typically preserved, although a great deal of transformation is taking place. For example, one might be able to grasp the total order of a symphony at a particular moment, thus suggesting that there is a timeless aspect to our musical experience (Bohm 1976, p. 41). Yet experiencing such a total order is not an entirely static experience but can involve an intense sense of movement.

Bohm next draws attention to an important difference between the ink-in-fluid model of the electron and musical experience:

> The key difference in these two cases is that for our model of the electron an enfolded order is grasped in thought, as the presence together of many different but interrelated degrees of transformations of ensembles, while for the music, it is sensed immediately as the presence together of many different but interrelated degrees of transformations of tones and sounds. In the latter, there is a feeling of both tension and harmony between the various co-present transformations of the music in its undivided state of flowing movement. (Bohm 1980, pp. 199–200)

There thus seem to be two ways in which a human being can comprehend an implicate order. There is what we might call *conceptual comprehension*, where the comprehension takes place with the help of a model which can be conceptually described and communicated. But there is also *non-conceptual comprehension*, where the comprehension happens via "immediate sensing" of the presence of sensory contents in conscious experience (cf. Pylkkö (1998)). Bohm further specifies the nature of this "immediate sensing":

> In listening to music, *one is therefore directly perceiving an implicate order.* (Bohm 1980, p. 200)

Implicate order, it seems, can thus be either "directly perceived" or "indirectly perceived". Of course, we can make a conceptual model about musical experience (e.g. the one Bohm has sketched above), and via that try to grasp the implicate order of musical experience conceptually, without actually listening to music. But we also have the option of putting on the CD player, sitting back, relaxing and directly perceiving the implicate order of our musical experience (or better still, pick up the guitar and play). Indeed this is a very powerful way to illustrate what is meant by conscious experience. It is thus perhaps not surprising that one of the leading analytical philosophers studying consciousness, David Chalmers, often starts his lecture by putting on the stereo and letting classical music illustrate what conscious experience is all about!

In contrast, we cannot (it seems) "directly perceive" the implicate order of the mode of existence and movement of an electron; our only way to grasp this implicate order is through the conceptual model (unless, say, it would somehow be possible for us to directly perceive or experience the implicate order of the mode of existence of the elementary particles that constitute our brains, just because in some sense we as conscious beings are (at least in part) constituted of a set of such particles, and more precisely their implicate orders).

Bohm next emphasizes yet another important aspect of the implicate order:

> Evidently this order is *active* in the sense that it continually flows into emotional, physical, and other responses that are inseparable from the transformations out of which it is essentially constituted. (Bohm 1980, p. 200)

Here he introduces the notion of *active order*, which clearly anticipates the notion of *active information*, which plays a central role in Bohm and Hiley's "ontological interpretation" of the quantum theory, which we shall discuss in more detail the next chapter. Such "active order" emphasizes the undivided wholeness of the mental and physical sides of a human being. The idea is that, say, the order of music – something that can be considered a fairly subtle, high-level mental property – can "flow into" emotional and physical responses. Bohm proposes that such responses cannot be separated from the transformations which constitute the order. What is going on here is not so much an "interaction" between mind and matter (conceived of as separate entities or "substances"), but rather a much more subtle relationship or mutual influence between them, which perhaps could be called "mutual participation". We shall return to discuss the nature of this relationship later on.

We have thus far been focusing on the question whether the implicate order prevails in musical experience. Of course, we could ask the same question about auditory experience in general. Consider, for example, someone speaking to you. You are taking in a whole number of "ensembles" (what we take as an "ensemble" is, of course, debatable; it could be a phoneme, a word, a sentence, or a set of sentences). Such ensembles are taken in by the auditory system somewhat analogously to the way we might put ink droplets into the glycerine. As time goes, on, many such ensembles are enfolded in the brain/mind. Thus we have a co-presence of (ensembles of) elements at different degrees of enfoldment – i.e. an implicate order! When we apprehend this implicate order, we sense it as meaning. If we are listening to a poem, we might also sense a kind of movement of meaning with emotional responses, not completely unlike the movement we grasp when listening to a symphony.

Auditory experiences thus illustrate the implicate order of conscious experience particularly well. Perhaps this is partly so because the auditory input often has a short duration in the explicate order. When listening to a symphony, only a tiny portion of the sounds is vibrating in the air at any particular moment; the rest is enfolded in your brain/mind.

In a typical visual experience, we meet a three-dimensional world of objects, which can be static. This strongly emphasizes the explicate order, and has perhaps encouraged many scientists and philosophers to assume that all that exists must in the end be nothing but a constellation of elements in the explicate order. But even visual experience involves the implicate order in a powerful way. Think again of language. When you read, you are a taking in "ensembles" (and again what we take as an "ensemble" is debatable; it could be a letter, a syllable, a word, a sentence, or a set of sentences). Recnet studeis suggset taht for unedrstadning writetn txet, what mattres is taht the fisrt and the lsat letetrs are corerct, wihle the odrer of the letetrs in bewteen deos not mkae scuh a big differnece. [*sic*] As you read these words, an implicate order is building up in a similar way to what takes place with spoken language or with listening to music.

The role of the implicate order in visual experience can also be vividly brought out by considering what takes place as we watch a motion picture, as Bohm points out:

> ... consider the sense of motion that arises when one is watching the cinema screen. What is actually happening is that a series of images, each slightly different, is being flashed on the screen. If the images are separated by long intervals of time, one does not get a feeling of continuous motion, but rather, one sees a series of disconnected images perhaps accompanied by a sense of jerkiness. If, however, the images are close enough together (say a hundredth of a second) one has a direct and immediate experience, as if from a continuously moving and flowing reality, undivided and without a break. (Bohm 1980, p. 200)

This is, of course, analogous to what happens with music. Here an "ensemble" (corresponding to the droplet in the ink-in-dye analogy) can be taken to be a single image. Such images are then being enfolded into the brain/mind, forming an implicate order (i.e. a set of co-present (ensembles of) elements at different degrees of enfoldment). The key point is that this gives rise to a "sense of motion". When watching a motion picture, there are a large number of images (e.g. one hundred per second). As is well known, the same feature or "illusion of movement" can be created by two flashing lights, with the aid of a stroboscopic device:

> Two disks, A and B, enclosed in a bulb, can be caused to give off light by means of electrical excitation. The light is made to flash on and off so rapidly that it appears to be continuous, but in each flash it is arranged that B will come on slightly later than A. What one actually feels is a sense of "flowing movement" between A and B, but paradoxically nothing is flowing out of B (contrary to what would be expected if there had been a real process of flow)... This means that a sense of flowing movement is experienced when, on the retina of the eye, there are two images in neighbouring positions one of which comes on slightly later than the other. (Bohm 1980, p. 200)

If you like, the two images on the retina are the "input", and the sense of flowing movement in conscious experience is the "output". Bohm emphasizes that the key factor making this output possible is the existence of an implicate order: a set of co-present (ensembles of) elements at different degrees of enfoldment in the brain/mind. The brain/mind is somehow sensing this implicate order, and this sensing gives rise to the sense of flowing movement. There are certain conditions that have to be satisfied, if an implicate order that can be sensed as movement is to build up in the brain/mind. For example, the interval between the stimuli (whether auditory or visual) must not be too large.

Bohm emphasizes that the sense of unbroken movement experienced when watching the motion picture or looking into the stroboscope is basically similar to that arising from a sequence of musical notes. One difference is that in music we can often resolve the notes in consciousness, while visual images typically arrive so close together in time that they cannot be thus resolved. Regardless of this difference, there is a basic similarity between auditory and musical experience.

> ...it is clear that visual images must also undergo active transformation as they "enfold" into the brain and nervous system (e.g., they give rise to emotional, physical, and other more subtle responses of which one may be only dimly conscious, as well as to "after images" that are in certain ways similar to the reverberations in musical notes). Even though the time difference of two such images may be small, the examples cited above make it clear that a sense of movement is experienced through the intermingling and inter-penetration of the co-present transformations to which these images must give rise, as they penetrate the brain and nervous system. (Bohm 1980, p. 201)

The above could provide an interesting tool for research in cognitive neuroscience. There is currently much research which tries to find the "neural correlates of consciousness" in the brain. A central aspect of conscious experience is a sense of movement, and Bohm's approach makes a hypothesis about the neural correlates of such experience. This could be summarized as a maxim: whenever there is a sense of movement in conscious experience, look for an implicate order in the brain. Thus, the neural correlate of experiencing the movement of a symphony is likely to involve an implicate order in the auditory cortex (and in other areas of the brain where this implicate order may have been transmitted). Likewise, the neural correlate of experiencing the apparent motion when watching the stroboscope is likely to involve an implicate order in the visual cortex (and in other associative areas of the brain where this implicate order may have been transmitted). More generally, our experience is multimodal, a fusion of information from different sensory modalities. When watching a motion picture, we are, for example, experiencing movement at many different levels. There is the movement of the objects in the screen, there is the movement of the music that may be played, there is the "movement of meaning" of the language spoken, and so on. The neural correlate of such experience thus involves many different implicate orders in the brain. The brain/mind is somehow able to sense all these together, giving rise to the overall multimodal conscious experience we typically have.

3.4 A Side-track: the Implicate Order and Zeno's Paradox

Bohm next moves on to consider how the implicate order might help us to understand motion in general (and not just the motion we sense in conscious experience):

> All of this suggests that quite generally (and not merely for the special case of listening to music), there is a basic similarity between the order of our immediate experience of movement and the implicate order as expressed in terms of our thought. We have in this way been brought to the possibility of a coherent mode of understanding the immediate experience of motion in terms of our thought (in effect thus resolving Zeno's paradox concerning motion). (Bohm 1980, p. 201)

What does he mean when saying that there is "a basic similarity between the order of our immediate experience of movement and the implicate order as expressed in terms of our thought"? Presumably this means that the order of, say, our experience when listening to music is basically similar to the order that prevails in the ink-in-fluid model when droplets have been enfolded there. Remember how Bohm earlier emphasized that the implicate order can be experienced in different ways. In the case of, say, music, we experience it directly. In the case of the electron, we experience it indirectly, in terms of our thought, for example with the help of the ink-in-fluid analogy, which can be described with concepts. What he emphasizes is that in his scheme there is a basic similarity between our direct experience of movement and our conceptual description of movement. Now, Zeno's paradox concerning motion has to do with the difference between our usual experience of motion and our usual way of thinking about motion in terms of our concepts. Our usual immediate sensory experience of motion is an object moving continuously across space. However, as Zeno made explicit, when we try to think of such motion, a paradox arises. It seems very natural to assume that an object (e.g. the tip of an arrow) occupies a single point in space at a single instant of time. When the object moves, it then goes to another point as time passes. One of the things Zeno pointed out, however, is that in order to get from a point

A to a point B, the object first has to go to a point C in between them, and that between A and C there is yet another point, D, and so on, without limit. This seems to make it impossible to understand how the object could ever move, because there are an infinite number of points the object should visit before it can reach any given point. If you like, in order to move, the object first has to go to the next, adjacent point, but it seems impossible to get there, and thus motion seems impossible.

Now, Bohm suggests above that he thinks he can resolve Zeno's paradox in the implicate order scheme. This would be no minor achievement, given that Zeno's paradox is one of the oldest philosophical problems of Western civilization. Let us therefore follow his argument in some detail, in order to be able to understand and evaluate his proposed solution. He starts by drawing attention to the way motion is usually thought of, in terms of a series of points along a line. If a particle moves, one typically assumes that at a given time t_1, a particle is at a position x_1, while at a later time t_2, it is at another position x_2. The velocity v of such a particle can be then be expressed as $(x_2 - x_1)/(t_2 - t_1)$. Bohm now starts his critique of this usual way of thinking (which in some key ways resembles Bertrand Russell's criticisms):

> Of course, this way of thinking does not in any way reflect or convey the immediate sense of motion that we may have at a given moment, for example, with a sequence of musical notes reverberating in consciousness (or in the visual perception of a speeding car). Rather, it is only an abstract symbolization of movement, having a relation to the actuality of motion, similar to that between a musical score and the actual experience of the music itself. If, as is commonly done, we take the above abstract symbolization as a faithful representation of the actuality of movement we become entangled in a series of confused and basically insoluble problems. (Bohm 1980, p. 201)

Bohm is thus strongly underlining the difference between our immediate sense of motion and the commonly used abstract symbolization of motion. If one takes the abstract symbolization as a representation of motion, one is led to assume that times t_1 and t_2 both exist. For presumably, if the abstract symbolization is assumed to represent something real, the things it represents must exist. The symbolization refers to t_1 and t_2 at the same time, so presumably t_1 and t_2 must exist at the same time. Bohm (1980, p. 202) emphasizes that the assumption that both t_1 and t_2 exist at the same time is in contradiction with our actual experience, which indicates that

> when a given moment, say t_2, is present and actual, an earlier moment, such as t_1, is past. That is to say, it is gone, non-existent, never to return.

The key trouble is that

> if we say that the velocity of a particular now (at t_2) is $(x_2 - x_1)/(t_2 - t_1)$ we are trying to relate what is (i.e., x_2 and t_2) to what is not (i.e., x_1 and t_1). (Bohm 1980, p. 202)

The usual view of time says that only the present and what is in it exists. The past is gone, the future is not yet. Yet the usual notion of velocity at a given now paradoxically involves both the present and the past.

Bohm admits that we can use the above expression as long as we remember that we are using it abstractly and symbolically, as, indeed, is commonly done in science and mathematics. But he emphasizes that the abstract symbolism cannot comprehend that

> the velocity *now* is active *now* (e.g., it determines how a particle will act from now on, in itself, and in relation to other particles). How are we to understand the *present* activity of a position (x_1) that is now non-existent and gone for ever? (Bohm 1980, p. 202)

He notes that it is commonly thought that this problem is resolved by the differential calculus.

> What is done here is to let the time interval, $\delta t = t_2 - t_1$ become vanishingly small, along with $\delta x = x_2 - x_1$. The velocity now is defined as the limit of the ratio $\delta x / \delta t$ as δt approaches zero. It is then implied that the problem described above no longer arises, because x_2 and x_1 are in effect taken at the same time. They may thus be present together and related in an activity that depends on both. (Bohm 1980, p. 202)

However, Bohm is not satisfied with this approach. He claims that

> this procedure is still as abstract and symbolic as was the original one in which the time interval was taken as finite. Thus one has no immediate experience of a time interval of zero length, nor can one see in terms of reflective thought what this could mean. (Bohm 1980, p. 202)

The procedure is thus both empirically and conceptually/logically ambiguous! Bohm thinks that when taken as an abstract formalism, the differential calculus is not fully consistent in a logical sense. Of course, it is widely used in contemporary science, so it does work up to a point. There is even a whole research programme in cognitive science, the "dynamical approach", which makes essential use of the differential calculus when describing various cognitive processes. However, Bohm emphasizes that the differential calculus has a limited range of applicability even for physical phenomena:

> ... it applies only within the area of continuous movements and then only as a technical algorithm that happens to be correct for this sort of movement. As we have seen, however, according to the quantum theory, movement is not fundamentally continuous. So even as an algorithm its current field of application is limited to theories expressed in terms of classical concepts (i.e., in the explicate order) in which it provides a good approximation for the purpose of calculating the movements of material objects. (Bohm 1980, p. 202–203)

The above point might also encourage us to reflect upon the applicability of dynamical systems theory (in so far as it relies on the different calculus) to describe cognition and consciousness. To be sure, cognition and conscious experience involve the sort of movement in the explicate order which can conveniently be described in terms of the differential calculus. However, one sees sometimes a tendency to assume that this approach works for cognition and conscious experience in a very broad and comprehensive sense (see, for example, van Gelder's (1999) attempt to discuss time consciousness in terms of the dynamical approach). I think that Bohm's above criticism of the differential calculus applies strongly in the case of time consciousness, thus suggesting that there are important limits of applicability to the dynamical approach in the domain of cognition and consciousness (see also Pylkkänen (forthcoming)).

Let us now move on to consider whether the implicate order can do any better than the dynamical approach/differential calculus as a description of movement:

> When we think of movement in terms of the implicate order, however, these problems do not arise. [*Footnote:* In the implicate order the basic algorithm is an algebra rather than the calculus.] In this order, movement is comprehended in terms of a series of inter-penetrating and intermingling elements in different degrees of enfoldment all present together. The activity of this movement then presents no difficulty, because it is an outcome of this whole enfolded order, and is determined by

> relationships of co-present elements, rather than by the relationships of elements that exist to others than no longer exist. (Bohm 1980, p. 203)

The key point is that an element (or an ensemble of elements) that has been enfolded still exists and is present. One way of interpreting the notion of enfoldment is thus that it extends our notion of what exists. The usual view of time says that only the explicate order at an instant t_1 exists at t_1; at a later instant t_2, the explicate order at t_1 no longer exists. In Bohm's ontology the explicate order at t_1 still exists at t_2, but in an enfolded state. The fact that it exists makes it intelligible that it can have an effect at t_2, which is not the case according to the usual view of time.

In this way, Bohm thinks he is able to find a coherent fit between abstract logical thought and concrete immediate experience:

> We see, then, that through thinking in terms of the implicate order, we come to a notion of movement that is logically coherent and that properly represents our immediate experience of movement. Thus the sharp break between abstract logical thought and concrete immediate experience, that has pervaded our culture for so long, need no longer be maintained. Rather, the possibility is created for an unbroken flowing movement from immediate experience to logical thought and back, and thus for an ending of this kind of fragmentation. (Bohm 1980, p. 203)

Not only does he think that the above helps to solve the problem of describing movement in terms of our thought, he also thinks it helps make sense of his proposed architecture of existence, namely the idea that reality is movement.

> In terms of the implicate order ... movement is a relationship of certain phases of what is to other phases of what is, that are in different stages of enfoldment. (Bohm 1980, p. 203)

To say that reality is movement is then to say that reality consists of different phases. The phases differ with regard to their degree of enfoldment. The explicate order at a given moment could be defined as that phase for which the degree of enfoldment is zero. The past could be defined as a set of phases for which the degree of enfoldment is larger than zero. Of course, this is straightforward to see if one considers the ink-in-fluid analogy. The droplet that is visible now has a zero as its degree of enfoldment. The degree for the other droplets can be defined in terms of how many times the outer cylinder has been turned since they were placed into the glycerine. When applied to reality, this model suggests that, at a given moment, there exists not only the explicate order, but also the orders that were previously explicate but have now become enfolded or implicate, as well as those enfolded orders or potentialities which have not yet become explicate. This suggests a new view about the essence of reality:

> ...the essence of reality as a whole is the...relationship among the various phases in different stages of enfoldment (rather than, for example, a relationship between various particles and fields that are all explicate and manifest). (Bohm 1980, p. 203)

Given that movement is assumed to be a co-presence of ensembles at different degrees of enfoldment, the above then amounts to saying that the essence of reality is movement.

We have thus seen how movement for Bohm implies the presence of an implicate order, that is, a set of co-present (ensembles of) elements at different degrees of enfoldment. We can directly experience such an implicate order (for example, in musical experience or when we are watching a motion picture), and when we do that we get an immediate intuitive sense of unbroken flow. But movement is not just the presence of an implicate order and the sense of flow; movement also involves change. Typically, when we experience a sense of flow, there will be a change in the state of affairs in the next moment. Bohm next considers whether this can also be understood in terms of the implicate order. He tackles the issue by considering the way the process of thought develops from moment to moment.

Bohm's discussion of thought, which follows, is interesting, but a few words of clarification might be useful before we proceed. In this chapter we have been closely following and discussing a section of the last chapter of *Wholeness and the Implicate Order* in which Bohm's aim is to consider whether the implicate order applies to consciousness. In doing that, he was led to consider Zeno's paradox of motion and the way the implicate order might resolve them. (I have called this extremely interesting discussion a "side-track".) Now, one could argue that Bohm was justified in doing this, because Zeno's paradox essentially involves conscious experience – after all, the paradox has to do with the fact that the sort of motion we experience in consciousness seems impossible in the light of our usual way of thinking about motion. His resolution of the paradox then involves proposing a new way to think about motion as essentially involving the implicate order, and a suggestion that this new way coheres naturally with our immediate experience of motion.

All this is profound, but it may also be somewhat confusing to the reader. To suddenly focus upon Zeno's paradox and problems with the differential calculus may appear to be a change of topic in a section (and in the case of the present book, a chapter) that was supposed to deal with consciousness and the implicate order. We also saw how Bohm, after describing motion in terms of the implicate order, raised the issue of how to understand the idea that reality is movement in terms of the implicate order, again "changing the topic", this time from consciousness to our general concept of reality. Finally, he draws attention to the fact that movement typically involves change, and sets out to explain how that is possible in the implicate order framework. At this point he returns to consider the mind, namely the nature of thought, but not, it seems for "its own sake" but in order to illuminate how to understand the nature of change in the implicate order framework. I think Bohm's argumentation is interesting, and mostly easy to follow. But one can also easily lose track of what he was primarily trying to do in this section. It seems he was trying to do a great many things!

Now, it is obvious that the implicate order is an extremely rich notion for Bohm. Because he thinks that it applies to consciousness and matter alike, it is perhaps understandable that he switches from one to the other fairly freely. However, I think that such switching makes this particular section fairly difficult. For example, I think that it is interesting to consider how the implicate order can be applied to describe the thought process for its own sake, and not just in so far as this helps us to understand the nature of change in terms of the implicate order. However, regardless of the above difficulties, I have chosen to follow the order of Bohm's argumentation in this chapter. I think the ideas he discusses are potentially ground-breaking and deserve careful consideration. I hope my discussion, as well as my consideration of some problems in his presentation, helps the reader to "divide and conquer" this difficult but interesting section in Chap. 7 of *Wholeness and the Implicate Order.*

3.5 The Implicate Order and the Process of Thought

Bohm begins his consideration of how the implicate order might apply to the process of thought by focusing on the way we use the word "imply" in the context of thought. What do we mean when, as we engage in thinking, we say that one set of ideas implies a different set?

> Of course the word "imply" has the same root as the word "implicate" and thus also involves the notion of enfoldment. Indeed,

> by saying that something is implicit we generally mean more than merely to say that this thing is an inference following from something else through the rules of logic. Rather, we usually mean that from many different ideas and notions (of some of which we are explicitly conscious) a new notion emerges that somehow brings all these together in a concrete and undivided whole. (Bohm 1980, p. 204)

Bohm goes on to describe thought and consciousness in more detail. He says that each moment of consciousness has a certain explicit content, which is a foreground, and an implicit content, which is a corresponding background. He then proposes that the process of thought is best understood in terms of the implicate order. He means not only the content of thought, to which it is fairly straightforward to apply the implicate order. More fundamentally, he suggests that that implicate order prevails in the "actual structure, function, and activity of thought". It is thus possible to distinguish between implicit and explicit aspects of thought. He proposes that this particular distinction is equivalent to the distinction between implicate and explicate in matter in general.

In the previous chapter we spent some time considering the causal architecture of the holomovement. Bohm's basic suggestion in this regard is that the laws that prevail in the various sub-totalities of the holomovement have the same basic form. The basic form, as we saw, is the following:

> ...the enfolded elements of a characteristic ensemble (e.g. of ink particles or of atoms) that are going to constitute the next stage of unfoldment are bound by a force of overall necessity, which brings them together, to contribute to a common end that emerges in the next phase of the process under discussion. (Bohm 1980, p. 204)

He proposes that the same causal architecture that prevails in inanimate matter and the biological domain also prevails in the domain of brain/mind. Thus, he postulates that

> the ensemble of elements enfolded in the brain and nervous system that are going to constitute the next stage of development of a line of thought are likewise bound through a force of overall necessity, which brings them together to contribute to the common notion that emerges in the next moment of consciousness. (Bohm 1980, p. 204)

It is one of Bohm's ambitions to show that matter (both inanimate and animate) and consciousness have in common the same basic order, the implicate order. If the implicate order prevails in a situation or a system, then one expects to find in it a "set of co-present (ensembles of) elements at different degrees of enfoldment". But the implicate order also essentially involves a causal architecture. The idea is that, in any given situation in which the implicate order prevails, there is a force of an overall necessity which brings the enfolded ensembles together to constitute an explicate end.[6]

Another idea that forms an important part of Bohm's view is that consciousness can be described in terms of a series of moments.

> Attention shows that a given moment [of consciousness] cannot be fixed exactly in relation to time (e.g., by the clock) but rather, that it covers some vaguely defined and somewhat variable extended period of duration. (Bohm 1980, p. 204)

Bohm is talking about "moments" as opposed to "events" of consciousness. This is analogous to the way he prefers to talk about moments as opposed to events at the fundamental level of physical ontology. In physics the point event is a concept emphasized by the theory of relativity. The uncertainty principle of quantum theory, especially when considered together with Einstein's general theory of relativity, suggests, however, that we have to give up the point event as fundamental and replace it by the notion of a series of moments which overlap with each other and which have a somewhat ambiguously or vaguely defined duration (Bohm 1986, p. 183). Bohm thus thinks that the moments of consciousness are analogous to the moments that make up reality in general.

Let us return to consider the moments of consciousness. Remember that one of Bohm's motivations for considering the process of thought is that in his view it provides a good way of understanding how change takes place. The key factor here is the force of overall necessity which brings enfolded elements in the brain/mind together to contribute to

[6] Bohm's approach can be seen as one way to try to achieve the unity of science, to seek a unified description of the phenomena studied by physics, biology, and psychology. The logical positivists likewise thought a certain kind of unity of science an important way, although they tried to achieve this unity in a different way. An interesting attempt toward a unity of science which deserves attention was developed by the Finnish philosopher Eino Kaila (1890–1958). The key concept in his approach was something he called "terminal causality". It is a principle (which, for example, emphasizes the role of boundary conditions) which he thought fundamental at the quantum level, and also in biological and psychological phenomena. See Kaila (1956, 1979).

the content that emerges in the next moment of consciousness. As this kind of process recurs, we can then understand how changes can take place from one moment to another.

> In principle, the change in any moment may be a fundamental and radical transformation. However, experience shows that in thought (as in matter in general) there is usually a great deal of recurrence and stability leading to the possibility of relatively independent sub-totalities. (Bohm 1980, p. 205)

This is a very important point. Bohm's implicate order scheme is fascinating, but it also has a terribly deterministic flavour. For example, if we assume that all the potentialities in the holomovement were fixed, then there would be no possibility for anything new to arise in the Bohmian universe. This is why it is important to remember that Bohm assumes that the holomovement is incomplete and in a constant process of unfoldment. Such unfoldment is not merely the actualization of pre-existing potentialities. This leaves room for genuine creativity and makes it plausible that the change in any moment may, at least in principle, be a "fundamental and radical transformation" (see also Bohm (1987c)). But note also that Bohm thinks that recurrence and stability have an extremely important role to play in the universe, for they make it possible for there to exist relatively independent sub-totalities. For example, we might think of an individual human being as such a "relatively independent sub-totality". Relative independence means that the individual has, in principle, some freedom. But the Bohmian scheme brings out how such freedom requires there to be recurrence and stability, properties that at first sight may seem the opposite of freedom. Recurrence and stability are required if the sub-totalities are to have any independence and autonomy in relation to the holomovement. Otherwise, the unending flux of the holomovement would destroy any independence, including the one required for freedom.

Within the subtotality, certain regularities then prevail:

> In any such sub-totality, there is the possibility of the continuity of a certain line of thought, that unfolds in a fairly regularly changing way. Evidently, the precise character of such a sequence of thoughts, as it unfolds from one moment to the next, will generally depend on the content of the implicate order in earlier moments. (Bohm 1980, p. 205)

This suggests that it might be reasonable to seek some models of the way a sequence of thoughts unfolds, which connects with the various computational approaches in cognitive science. Indeed, artificial neural networks have some properties that seem to fit very well to describe the way Bohm thinks (part of) the implicate order at a given moment unfolds to the explicate order at a next moment. Artificial neural networks can be described with the same mathematics that can be used to describe quantum processes (see, for example, Perus (1995)). Thus the implicate order approach does not exclude the use of computational tools to describe cognition and conscious experience. On the contrary, it suggests interesting new ways to do this (for example, in terms of an algebra which takes unfoldment and enfoldment as fundamental, see Bohm (1980, pp. 157–171).)

3.6 The Role of the Explicate Order in Conscious Experience

We have now seen that Bohm thinks that the implicate order prevails in various ways in the brain/mind and in conscious experience. Memory may be stored in the brain in a somewhat holographic fashion, suggesting that the implicate order prevails in the neural processes correlated with mental processes and conscious experience. The implicate order also prevails in conscious experience whenever we experience movement, be it the movement of a symphony or watching a moving object. Finally, the implicate order seems to prevail in the way our thoughts unfold from moment to moment. Thus, there seems to be a fairly strong case for the suggestion that the notion of implicate order can be applied to consciousness, thus opening up the possibility of realizing Bohm's vision that the implicate order could be extended into a theory about the architecture of existence in general. However, the world we typically meet in everyday experience is not the sort of "unbroken wholeness in flowing movement" Bohm thinks fundamental, but rather something we might, tongue-in-cheek, call "broken partiality at rest". This, of course, is the explicate order, and Bohm admits that the explicate order is usually the order we perceive through the senses. Indeed, the explicate order could almost be *defined* as the world that we meet through the senses. One explanation for this might be that our senses and measuring instruments can only register stable and manifest aspects of the world, and that is what the explicate order, by definition, is (Bohm 1980, p. 186). In contrast, the holomovement is so unstable and changing that it escapes a typical empirical attempt to capture it. This is how Bohm starts to unpack the question of why is it that we typically find the explicate order to be the "manifest content" of our conscious experience:

> As observation and attention show (keeping in mind that the word "manifest" means that which is recurrent, stable, and separable), the manifest content of consciousness is based essentially on memory, which is what allows such content to be held in a fairly constant form. Of course, to make possible such constancy it is also necessary that this content be organized, not only through relatively fixed associations but also with the aid of the rules of logic, and of our basic categories of space, time, causality, universality, etc. In this way an overall system of concepts and mental images may be developed, which is a more or less faithful representation of the "manifest world". (Bohm 1980, p. 205)

Here we see again the way Bohm incorporates a "Kantian epistemology" into his implicate order approach. For Bohm does not think that our conscious experience is a direct "copy" of the external world. On the contrary, conscious experience involves a great deal of construction and ordering on the part of the brain/mind, where the basic categories reside. When we point to the similarity between Kant and Bohm in this respect, it is again important to note that Kant did not seek to naturalize his forms of perception (space and time) and categories of understanding (e.g. causality) by assuming that they are neurally based. Bohm from his 20th-century perspective can go further along with such naturalization, but as we will see, neither did he think that conscious experience can be entirely reduced to neural processes. Instead, both conscious experience and neural processes are in the end abstractions or "correlated projections" from the holomovement, in which they meet. Thus, they are not really separate because they are ultimately one at the level of the holomovement. Further, it is futile to try to reduce conscious experience entirely to neural processes, because neural processes do not have the status of an autonomous substance which would be required for them to be able, on their own, to produce consciousness. Thus, the apparent difference between Kant and Bohm on the question of naturalization need not be that great. Kant did not think that the forms of perception and the categories of understanding are neural processes; but nor, it seems, did Bohm, strictly speaking. Bohm would certainly agree that neural descriptions can illuminate a great deal the role of, for example, space, time, and causality in conscious experience. However, he did not assume that the essence of conscious experience was in the neural processes; this essence, like the essence of everything, ought to be sought in the holomovement, the ground of existence as we think of it today. For the above reasons, I often prefer to talk about the "brain/mind", as opposed to just "brain" or "mind" (cf. P. S. Churchland (1986)). The idea of the solidus is to emphasize the ultimate unity of the brain and the mind, but the use of the two words is also a reminder that mind cannot be entirely reduced to the brain, but is a different and relatively autonomous aspect of the holomovement.

Bohm again emphasizes that thought (presumably used here in the broad sense to include "conscious experience") is not just a repetition or "re-presentation" of what is already "present" in external reality "out there".

> The process of thought is not, however, merely a *representation* of the manifest world; rather, it makes an important *contribution* to how we experience this world, for ... this experience is a fusion of sensory information with the "replay" of some of the content of memory (which latter contains thought built into its very form and order). (Bohm 1980, p. 205)

We already briefly discussed such "fusion" above when considering Pribram's holographic theory of memory. If both incoming sensory information and the information contained in memory exist as patterns of neural activity, this makes it possible, in principle, to understand how this fusion could take place. The exact way in which such fusion happens is, of course, very complex, but it is conceivable that it can happen, given that both incoming information and information in the memory exist as patterns of nervous energy.

> In such experience, there will be a strong background of recurrent stable, and separable, features, against which the transitory and changing aspects of the unbroken flow of experience will be seen as fleeting impressions that tend to be arranged and ordered mainly in terms of the vast totality of the relatively static and fragmented content of recordings from the past. One can, in fact, adduce a considerable amount of scientific evidence showing how much of our conscious experience is a construction based on memory organized through thought, in the general way described above. (Bohm 1980, pp. 205–206)

Bohm discussed such evidence in an illuminating way in the appendix "Physics and perception" to his 1965 book *The Special Theory of Relativity*, where he also briefly discussed his views about Kant. Basically, Bohm thinks that Kant was correct in emphasizing the way the human mind constructs experience, but mistaken in attributing his forms of perception and categories of understanding a fixed quality. In this appendix, Bohm also discussed Piaget's and Gibson's views, among others, which inspired him a great deal.

The above idea of "a strong background of recurrent stable, and separable, features" reflects the idea that conscious experience essentially involves a kind of "virtual reality" associated with the brain (an idea advocated in recent consciousness studies by Velmans, Revonsuo, and Metzinger, for example). In the above-mentioned appendix, Bohm uses the term "inner show" to refer to this virtual reality:

> ... we do not actually create the world. In fact, we only create an "inner show" of the world in response to our movements and sensations. (Bohm 1965c, p. 216)

There is a disagreement among philosophers about whether perception involves a virtual reality or whether we in some sense perceive the external world directly. As especially Antti Revonsuo has emphasized, dreams provide strong support for the idea of conscious experience involving a virtual reality or (in Bohm's terms) a constructed inner show (Revonsuo 1995). In a dream, I typically find myself in a world, having various kinds of sensory experiences. However, when dreaming, I am not actually perceiving the external world in the usual way, so surely the brain/mind must have constructed the inner show of the world I am experiencing.[7]

Bohm is concerned with understanding how this virtual reality or inner show (which is a kind of explicate order) arises from the implicate order. He also raises the question of whether the explicate order typically dominates conscious experience too much, making us blind to the more implicate, transitory, and changing aspects of experience. He

[7] The psychiatrist and philosopher Gordon Globus has also developed the idea that resembles that of consciousness as virtual reality in his 1987 book *Dream Life, Wake Life*. Consider, for example, the following: "This life-world [during dreaming] is not created by a 'syntactical' rearranging of mnemic copies, as Freud would have it, but is instead 'formatively' created *de novo* by our own mental acts ... dreaming and waking lives are indistinguishable as *unreflectively lived lives* ... Both lives demand the same explanation on grounds of parsimony and evolutionary selection, and so if the dream life is 'formatively created', then so is the wake life. We think up the very lifeworld of waking too. We are enclosed within a 'bubble of perception' of our own making, as Castañeda's sorcerer says. Our true condition is monadic." (Globus 1987, p. viii). Globus also comments explicitly on Bohm's views: "My basic idea is that the possible worlds do 'exist', but as an enfolded order ... Possibility is implicate existence. Actuality depends on a process of unfolding enfolded order to explicate existence." (Globus 1987, p. 136). For further passages relevant to the idea of virtual reality, see Globus (1987, pp. 106, 172, 179).

thinks that Piaget's work is particularly important if we want to understand how the explicate and implicate orders of conscious experience are related:

> Piaget has made it clear that a consciousness of what to us is the familiar order of space, time, causality, etc. (which is essentially what we have been calling the explicate order) operates only to a small extent in the earliest phases of life of the human individual. Rather ... for the most part infants learn this content first in the area of sensori-motor experience, and later as they grow older they connect such experience with its expression in language and logic. On the other hand, there seems to be an immediate awareness of movement from the very earliest. Recalling that movement is sensed primarily in the implicate order, we see that Piaget's work supports the notion that the experiencing of the implicate order is fundamentally much more immediate and direct than is that of the explicate order, which, as we have pointed out above, requires a complex construction that has to be learned. (Bohm 1980, p. 206)

Bohm thus appeals to developmental psychology to further back up his proposal that the implicate order is the fundamental feature of conscious experience, while the explicate order is derivative and secondary.

Now, if the implicate order is primary in conscious experience, why do we then not notice it? One reason, Bohm suggests, is that we have become habituated to the explicate order. This connects with his idea that the explicate order has over the centuries become the dominant order in Western philosophical and scientific thinking. As a result, our thought and language is "biased" in favour of the explicate order, giving rise to a strong feeling that our primary experience is of that which is explicate and manifest (Bohm 1980, p. 206).

He also points to the nature of memory as an important reason for why the explicate order tends to dominate conscious experience. The content of memory recordings is mainly that which is "recurrent, stable, and separable". Once such recordings are activated (which, of course, is the case in a typical experience), our attention tends to be focused on what is "static and fragmented":

> This then contributes to the formation of an experience in which these static and fragmented features are often so intense that the more transitory and subtle features of the unbroken flow (e.g., the "tranformations" of musical notes) generally tend to

> pale into such seeming insignificance that one is, at best, only dimly conscious of them. (Bohm 1980, p. 206)

So, according to Bohm it is part of the "human condition" that although the implicate order (with features such as wholeness and flow) is fundamental and primary, we are almost doomed to an experience in which the static and fragmented features dominate. Of course, it is common practice to refer to experience (especially sensory experience) in science and philosophy as an arbiter of what is correct. Bohm, however, is suggesting that the very experience we are using to judge whether something is illusion or reality, correct or incorrect, etc. is itself biased in a very serious way. He writes:

> Thus, an illusion may arise in which the manifest static and fragmented content of consciousness is experienced as the very basis of reality and from this illusion one may apparently obtain a proof of the correctness of that mode of thought in which this content is taken to be fundamental. [*Footnote:* This illusion is essentially the one... in which the whole existence is seen as constituted of basically static fragments.] (Bohm 1980, p. 206)

So what Bohm is suggesting is that certain contents of consciousness are taken too seriously and generalized so that they give rise to an illusion about the nature of reality. This illusion is, in turn, used as evidence for the correctness of the way of interpreting the contents (that is, assuming they reflect the essence of reality), which gives rise to the illusion in the first place. If Bohm is correct, we are thus caught in a circular kind of self-deception. It is common in science to think that as long as we rely on sense perception, we can safely avoid illusion. Bohm calls for caution here, suggesting that sense perception is partly a function or outcome of contents stored in memory, which in turn may be biased and overly static and fragmented.

This is somewhat reminiscent of the way philosophers of science (e.g. Hanson, Kuhn, Feyerabend) have emphasized that observation, including scientific observation is "theory-laden". But Bohm does not here talk so much about a situation in which a scientific theory affects the way in which we set up the experiments and interpret the observations. Instead he is drawing attention to how the very general contents that have been abstracted from our experience and are stored in our memory have become the basis on which we interpret and experience the fundamental nature of the world. These contents, reflecting the explicate order, represent the world as static and fragmented. When we take them to be a description of the fundamental nature of the world, we are then led to experience existence as a whole as static and fragmented. According to him, this experience is an illusion, and it is very difficult to become free from it. Note that the experience that the world has a static aspect, consisting of relatively separate and independent objects, is not, as such, an illusion. The illusion is the experience that this is all there is to our existence. And because it seems to be our immediate experience (rather than our conscious reasoning) which is telling us this, it is, of course, very difficult to see that it is an illusion that is taking place. But, of course, the whole point of an illusion is that one does not see it as an illusion. Bohm's suggestion, if correct, would have far-reaching consequences for science, philosophy, and everyday life. Science and philosophy pay a lot of attention to securing that our knowledge is correct. However, Bohm is suggesting that much of science and philosophy is based on the same illusion as human experience more generally, namely the illusion that reality is fundamentally static and fragmented. Of course, he is not alone in suggesting this. For example, philosophers of a "post-phenomenological" type have made similar suggestions (see, for example, Pylkkö (1998), Globus (2003)).[8]

3.7 Matter, Consciousness, and the Architecture of Existence

While going through Bohm's presentation of his view, we have seen a certain view of matter, consciousness, and their relationship emerge. Our dualistic tradition saw matter and consciousness as two independent substances; our present materialistic age tries, in one way or the other (for example, via reduction, emergence, causation, and supervenience), to incorporate consciousness into the material realm. Bohm disagrees with both substance dualism and materialism. Perhaps his main contribution to the mind–matter debate is to show that the sort of notion of matter which philosophers typically presuppose and take as fundamental is an illusion. It is not an illusion to say that the universe has an aspect which consists of particles and fields in mechanical interaction. But empirical science seems to suggest that it is an illusion

[8] In my view, one of the most interesting and original philosophical books to have appeared in recent years is Pauli Pylkkö's *The Aconceptual Mind. Heideggerian Themes in Holistic Naturalism* (Pylkkö 1998). Pylkkö underlines the fundamentally aconceptual nature of the mind, and his views have some interesting similarities to Bohm's views above, although there are also important differences. See also Pylkkänen (2004b) and Globus (2003).

to think this aspect is the fundamental substance of the universe, into which everything else has to be reduced or incorporated.

We have also seen the general outlines of Bohm's own way of thinking about matter, consciousness, and their relationship. His basic proposal is that what is fundamental, autonomous, and self-existent is the holomovement, and that inanimate matter, animate matter, and consciousness are relatively autonomous sub-totalities that can be abstracted from the holomovement. They are not fragmented from each other because they "meet" or are the same at the level of the holomovement. They cannot be reduced to each other, because none of them has the status of a substance or something to which other aspects can be entirely reduced. Of course, Bohm's view has predecessors in the history of Western philosophy. There is some resemblance to Aristotle's ideas about form and matter being the two aspects of a substance. There is some resemblance to Spinoza's view of matter and consciousness as two aspects of an underlying reality, as well as later versions of "neutral monism", such as Russell's. Further, Bohm's emphasis on the implicate order brings to mind Leibniz's monads, the entities which reflect the entire universe. Bohm's description of non-locality as a "non-causal relationship" brings to mind Leibniz's idea that the monads are "windowless". The monads are not in causal interaction with each other, and yet they together constitute an ordered world. Leibniz refers to the idea of pre-established harmony to account for this, and also assumes that such pre-established harmony prevails between the mind and the body.[9] Assuming that the implicate order and especially non-locality are the most radical new features of the quantum theory, I find it amazing that Leibniz had anticipated these well before quantum phenomena were empirically discovered. Perhaps this can be seen as evidence about the "systemic character" of the universe. One can come to at least part of the essence of a fundamental theory of matter (e.g. quantum theory) before the empirical phenomena leading us to formulate that theory even have been discovered!

Let us now continue to follow Bohm's presentation of his implicate order theory. When discussing consciousness, he further developed the notion of the implicate order (for example, by introducing the idea the conscious experience consists in a series of overlapping "moments"). His

[9] It is important to remember here that for Leibniz mind and body are not two different substances. Physical objects, including the body, have the status of appearances, while the mental entities, the monads, are the primary substances. Thus the pre-established harmony between the mind and the body is a harmony between a substance (monad) and an appearance (the physical body). See McCann (1994).

challenge now is to show that this developed notion of the implicate order can be applied to matter, thus making stronger the idea that the implicate order prevails in both matter and consciousness. For Bohm, the possibility that the same order prevails in both suggests that they may have a common ground. This then gives rise to the view that matter and consciousness are two different aspects of an underlying reality.

He begins by suggesting that, just as with consciousness, in physics, too, we ought to take the moment (instead of the point event) as fundamental:

> Current relativistic theories in physics describe the whole of reality in terms of a process whose ultimate element is a point event, i.e. something happening in a relatively small region of space and time. We propose instead that the basic element be a moment which, like the moment of consciousness, cannot be precisely related to measurements of space and time, but rather covers a somewhat vaguely defined region which is extended in space and has duration in time. (Bohm 1980, p. 207)

It seems that Bohm is using an idea derived from the psychological domain, the "moment of consciousness", as a prototype for a notion of moment that applies to reality more generally. The more scientific reason for giving up the notion of point event in favour of an ambiguous moment has to do with the indeterminacy relations of quantum theory (see Bohm (1986, p. 183)). At least from the point of view of current empirical evidence, quantum theory does not support the plausibility of the point event as fundamental.

He next comments on how a moment is to be defined:

> The extent and duration of a moment may vary from something very small to something very large, according to the context under discussion (even a particular century may be a "moment" in the history of mankind). (Bohm 1980, p. 207)

For a long time this passage bothered me. On the one hand, Bohm is trying to provide us with an ontology, a description of the world as it exists independently of us. He tells us that the basic unit of his ontology (both the ontology of conscious experience and of reality more generally) is that of a moment. But the above quotation seems to suggest that the extent and duration of a moment is a matter of choice – it depends on us. Well, at least that's the way I read this passage for a long time, finding it fairly frustrating, feeling that it makes Bohm's ontology far too observer-relative by his own standards.

However, I recently thought of another, more objectivist way of interpreting the statement. For Bohm does not, strictly speaking, say that the extent and duration of a moment is a matter of our choice (for instance, that it is a question of whether I choose to consider a century or a day as the duration of a moment). He is only, it seems, saying that it may vary according to the context under discussion. Thus, even if I consider a particular century the "moment", this does not make the duration of this moment a subjective feature, something that exists only because I attribute it to reality. Rather, the fact that a particular century could be seen as a moment in the history of humankind can reflect the structure of reality itself. Thus, I think Bohm's idea is that reality itself has (as relatively autonomous aspects) various contexts in it, and that in each context we will find a certain structure of overlapping moments.

Of course, there remains the fundamental philosophical question to what extent our ideas about the moments correspond to the putative actual moments in the structure of reality. Bohm is somewhat ambiguous about how we should interpret the ontological status of such moments. He is often keen to say that such features of reality are "abstractions", implying that we abstract them conceptually. At the same time, he seems to think that there is something in the objective reality "out there" which makes it reasonable to abstract some things and not others. Such tension between the "subjective" and the "objective" is typical of the traditional philosophical discussion of this issue. For a more detailed discussion of this tension in Bohm's views, see the debate between Schindler (1982) and Bohm (1982).

For our purposes, the main point is that even though Bohm's description of moments may seem to make the moments a very observer-relative affair, the moments may have a more objective existence as relatively autonomous sub-totalities, for example an existence similar to that of chairs, say. We abstract chairs out of reality; at the same time there is something in reality which makes the abstraction reasonable. Someone might note here that chairs are observer-relative things in the sense that they exist relative to the intentions of the observer (e.g. the intention to sit) (see, for example, Searle's (1992) distinction between "intrinsic" and "observer-relative" properties). However, according to Bohm, the same applies even to more "intrinsic properties" of reality such as atoms. The physicist abstracts the atoms out of reality; and yet there is something in reality which makes this abstraction a reasonable one. The difference between chairs and atoms is that chairs exist as chairs primarily relative to our intention to use them, while atoms exist as atoms relative to the architecture of reality (for example, atoms make the stability of matter possible).

Let us move on. Bohm next brings out further ways in which the structure of reality is similar to the structure of consciousness:

> As with consciousness, each moment has a certain explicate order, and in addition it enfolds all the others, though in its own way. So the relationship of each moment in the whole to all the others is implied by its total content: the way in which it "holds" all the others enfolded within it. (Bohm 1980, p. 207)

This suggests that something like "memory" has a very powerful metaphysical significance in Bohm's view. Each moment "remembers" the previous moments, and in this sense each moment enfolds the whole universe. Note that this emphasizes time. Enfoldment does not only mean that information about the whole universe is enfolded in a given region of space (e.g. in the movement of electromagnetic waves). It also means that moments that no longer exist are in some sense present in the present moment. Of course, something like this is captured by the example of the electromagnetic waves. Sometimes we "see" a star that no longer exists. It may take such a long time for light to travel to the Earth from the star that by the time the light hits the retina of someone's eye, the star no longer exists. This is a vivid example of how one, existing moment can enfold information about another, non-existing moment.

Bohm himself notes that his idea that reality consists of moments which enfold each other resembles Leibniz's metaphysics:

> In certain ways this notion is similar to Leibniz's idea of monads, each of which "mirrors" the whole in its own way, some in great detail and others rather vaguely. The difference is that Leibniz's monads had a permanent existence, whereas our basic elements are only moments and are thus not permanent. (Bohm 1980, p. 207)

The expression "only moments" is very important. It suggests that the building blocks of the Bohmian universe have a fairly weak ontological status. The world consists of elements that are not permanent. Of course, if one's fundamental ontological assumption is that "all is flux", then something like that has to apply to the building blocks of existence.

Bohm also briefly comments on the relation between his and Whitehead's views:

> Whitehead's idea of "actual occasions" is closer to the one proposed here, the main difference being that we use the implicate order to express the qualities and relationships of our moments, whereas Whitehead does this in a rather different way. (Bohm 1980, p. 207)

Whitehead, of course, is one of the relatively few modern philosophers who, like Bohm, tried to formulate a general view of reality on the basis of modern physics. I think Bohm had an advantage here because he worked in a period during which quantum theory was developed (and indeed contributed himself significantly to the application and interpretation of quantum theory). At the same time, Whitehead's strong background in mathematics and philosophy (and his contact with Russell, for example) makes his philosophical views more comprehensive than those of Bohm.

Many Bohmian notions find their equivalent in Whitehead's philosophy (for example, unfoldment and enfoldment are very similar to Whitehead's concresence and transjection, and as we saw above, Bohm notes that his notion of moment is similar to Whitehead's idea of "actual occasion"). A basic similarity is, of course, also the more general assumption that reality is to be understood as a process. Bohm himself discusses the relationship of his views to those of Whitehead in his article in the 1986 anthology *Physics and the Ultimate Significance of Time: Bohm, Prigogine, and Process Philosophy* (Bohm 1986).

Bohm's next challenge is to connect the explicate order of conscious experience with the explicate order of reality in general. He has noted before that the explicate order of reality seems to be the same as the order revealed in sensory experience. He admits, however, that he has not explained why this is so, and proceeds therefore to try to give an explanation:

> We now recall that the laws of the implicate order are such that there is a relatively independent, recurrent, stable sub-totality which constitutes the explicate order, and which, of course, is basically the order that we commonly contact in common experience (extended in certain ways by our scientific instruments). (Bohm 1980, p. 207)

One important feature of the explicate order is that it makes memory possible, or the storing of information about previous moments into a given moment:

> This [explicate] order has room in it for something like memory, in the sense that previous moments generally leave a trace

> (usually enfolded) that continues in later moments, though this trace may change and transform almost without limit. (Bohm 1980, pp. 207–208)

Remember that the building blocks of the Bohmian universe are moments, and are thus not permanent. Yet the essential idea of his ontology is the idea of the implicate order: each moment contains within it all the other moments in some way. It is the stability of the explicate order which makes such enfoldment, and a kind of "presence of the past", possible.

According to Bohm, it is a very fundamental feature of the mode of being of the universe that it consists of moments. A moment has a limited duration as a moment; but it can continue its existence, at least in some sense, as an enfoldment or trace in later moments. If you like, there is a kind of "parasitic" quality to the architecture of the Bohmian universe. The past moments are parasites on the present moment. They can no longer exist independently as moments, so they exist as traces in the present moments. The explicate order thus provides a kind of continuity of existence and dependence of the present moment upon the past moments in a universe which otherwise would consist of moments ignorant of other moments. In this sense the explicate order makes the implicate order possible. Without the explicate order there would be no way for the past moments to exist in the present moment, and the world would lose its structure.

Bohm continues:

> From this trace (e.g., in the rocks) it is in principle possible for us to unfold an image of past moments, similar in certain ways, to what actually happened; and by taking advantage of such traces, we design instruments such as photographic cameras, tape recorders, and computer memories, which are able to register actual moments in such a way that much more of the content of what has happened can be made directly and immediately accessible to us, than is generally possible from natural traces alone. (Bohm 1980, p. 208)

The above suggests that various media have a certain "metaphysical function". The media enable the past moments to exist and exert influence in the present moment. They play a role in making the implicate order architecture of existence possible. Bohm next points out that our memory plays a similar role:

> One may indeed say that our memory is a special case of the process described above, for all that is recorded is held enfolded

> within the brain cells and these are part of matter in general. The recurrence and stability of our own memory as a relatively independent subtotality is thus brought about as part of the very same process that sustains the recurrence and stability in the manifest order of matter in general. (Bohm 1980, p. 208)

The role of our memory is thus to allow the past moments of our existence to be present in the present moment. Bohm now thinks he can explain why the order we typically meet in sensory experience is the explicate order of the physical universe:

> It follows, then, that the explicate and manifest order of consciousness is not ultimately distinct from that of matter in general. Fundamentally these are essentially different aspects of the one overall order. This explains a basic fact... that the explicate order of matter in general is also in essence the sensuous explicate order that is presented in consciousness in ordinary experience. (Bohm 1980, p. 208)

Bohm moves on to say that in many important respects

> consciousness and matter in general are basically the same order (i.e., the implicate order as a whole). (Bohm 1980, p. 208)

This expression illuminates Bohm's thinking fairly neatly. The idea is that there exists an "implicate order as a whole" which contains both consciousness and matter. Presumably "consciousness" refers to a certain aspect of this total order, and "matter" to another aspect. But the key point is that they are aspects of a single order. This, according to Bohm, makes their relationship possible. But how? This is what he sets out to discuss next:

> We may begin by considering the individual human being as a relatively independent sub-totality, with a sufficient recurrence and stability of his total process (e.g., physical, chemical, neurological, mental, etc.) to enable him to subsist over a certain period of time. (Bohm 1980, p. 208)

This is straightforward. We already noted above how recurrence and stability make it possible for sub-totalities to exist. In this sense recurrence and stability, although apparently features which are the opposite of freedom, actually seem to be necessary preconditions for freedom. For without them, a human being (and everything else) would be determined by the holomovement, and would not be able to perform any autonomous actions.

Bohm then begins to discuss the relationship between matter and consciousness:

> In this process we know it to be a fact that the physical state can affect the content of consciousness in many ways. (The simplest case is that we can become conscious of neural excitations as sensations). Vice versa, we know that the content of consciousness can affect the physical state (e.g. from a conscious intention nerves may be excited, muscles may move, the heartbeat change, along with alterations of glandular activity, blood chemistry, etc.). (Bohm 1980, p. 208)

So he admits that there is a sense in which matter affects consciousness and consciousness affects matter. However, he wants to question the traditional way of thinking about such influence:

> This connection of the mind and body has commonly been called psychosomatic (from the Greek "psyche", meaning "mind", and "soma", meaning "body"). This word is generally used, however, in such a way as to imply that mind and body are separately existent but connected by some sort of interaction.
>
> Such a meaning is not compatible with the implicate order. In the implicate order we have to say that mind enfolds matter in general and therefore the body in particular. Similarly, the body enfolds not only the mind but also in some sense the entire material universe. (In the manner explained earlier ... both through the senses and through the fact that the constituent atoms of the body are actually structures that are enfolded in principle throughout all space.) (Bohm 1980, p. 209)

Remember that we have previously mentioned a number of different relations that philosophers have suggested hold between matter and consciousness, e.g. reduction, emergence, causation and supervenience. It seems that Bohm suggests yet another relation: enfoldment. He says that mind enfolds matter in general and therefore the body in particular. What does this mean? Presumably this means that what we typically call matter has mind in some sense enfolded within it, at least as a potentiality. The philosopher Thomas Nagel has drawn attention to a similar feature in his article "Panpsychism" (Nagel 1987). He points to the fact that it is quite common in contemporary science and philosophy to assume that the properties of a whole system in some way derive from the properties of the parts. Thus no "radical emergence" is allowed, i.e. an assumption that some new property can emerge in a system, with no possibility of explaining how it emerged. If we now accept that conscious experience is a real property of the brain, we have dilemma. It seems that conscious experience in some way has to derive from the properties of the parts of the brain (in the end from the constituent elements, atoms, that make up the brain). On the other hand, our scientific tradition strongly suggests that atoms are purely physical, having no properties whatsoever that could explain the origin of conscious experience. Nagel reminds us that typical contemporary materialism allows that any suitable "non-living" and "non-conscious" atoms (e.g. those in a field where cows eat the grass, which in turn becomes the milk we drink), when suitably arranged, give rise to a system which is conscious. He thus notes that some sort of panpsychism is implicit in all those materialistic views which do not allow radical emergence and accepts that conscious experience is a real property. Nagel does not endorse panpsychism but sees it rather as a symptom that we might need to change our typical assumptions about the origin of consciousness.

Bohm's above suggestions are extremely interesting when considered in relation to Nagel's discussion. For Bohm's notion of "enfoldment" might be just the sort of notion we need to explain how "inanimate matter" can be combined to make a "conscious organism", without appealing to a mysterious, something-out-of-nothing radical emergence. If mind is, in some sense, enfolded in matter in general, this provides a qualitative explanation for why material systems can be conscious. Of course, it also opens up the possibility of "artificial mind", provided we arrange matter in such a way as to allow the enfolded mental (and perhaps proto-conscious) properties to actualize (see also Pylkkänen (1996a)).

Bohm next raises a very interesting and radical possibility: the relationship between matter and consciousness is similar to the relationship between quantum systems that are non-locally correlated. In connection with quantum non-locality, he introduced, as we saw above, the notion of

> a higher-dimensional reality, which projects into lower-dimensional elements that have not only a non-local and non-causal relationship but also just the sort of mutual enfoldment that we have suggested for mind and body. (Bohm 1980, p. 209)

He thinks that mind and body similarly might involve a higher-dimensional reality:

> ... we are led to propose further that the more comprehensive, deeper, and more inward actuality is neither mind nor body but

> rather a yet higher-dimensional actuality, which is their common ground and which is of a nature beyond both. (Bohm 1980, p. 209)

As we have noted above, this view in some ways resembles neutral monism. The reference to "higher-dimensional actuality" begins to show the way this view is inspired by the quantum theory.

> Each of these [i.e. mind and body] is then only a relatively independent sub-totality and it is implied that this relative independence derives from the higher-dimensional ground in which mind and body are ultimately one (rather as we find that the relative independence of the manifest order derives from the ground of the implicate order). (Bohm 1980, p. 209)

This makes clear the ontological status of mind and body. They are not substances, i.e. they are not autonomous and self-standing. Instead they are merely relatively independent sub-totalities. They owe this relative independence to the ground and the principles that prevail in this ground. Strictly speaking, mind and body are not separate, but "ultimately one".

Bohm next points out that

> [i]n this higher-dimensional ground the implicate order prevails. (Bohm 1980, p. 209)

Remember that the implicate order and movement are essentially related. This means that

> within this ground, what is, is movement which is represented in thought as the co-presence of many phases of the implicate order. (Bohm 1980, p. 209)

So, on the one hand there is movement, and on the other hand there is the possibility of representing this movement in thought in terms of the idea of a "co-presence of many phases of the implicate order". This, of course, is the idea of movement we already saw above. Conscious experience involves a sense of flow, indeed referred to by such metaphors as the "stream of consciousness". Bohm's idea is that such flow exists at the level of the ground. He next considers how we are to understand the role of change in this movement. Of course, it is our experience that the state of mind and the state of body change from moment to moment. But how does he account for this in the implicate order scheme?

> As happens with the simpler forms of the implicate order considered earlier, the state of movement at one moment unfolds through a more inward force of necessity inherent in this overall state of affairs, to give rise to a new state of affairs in the next moment. (Bohm 1980, p. 209)

The idea here is that the state of movement at a given moment involves an implicate or enfolded order. Because there is a tendency for enfolded states to unfold in any Bohmian "sub-totality", we expect things to be different in the next moment. The previously enfolded state has unfolded, and what was previously unfolded has enfolded. Of course, there can be recurrence and stability in such a process. But there is also the possibility of change, depending on the enfolded potentialities at any given moment.

Bohm next considers how we are to understand in his scheme the common fact that the state of the mind and the state of the body are correlated from moment to moment:

> The projections of the higher-dimensional ground, as mind and body, will in the later moment both be different from what they were in the earlier moment, though these differences will of course be related. So we do not say that mind and body causally affect each other, but rather that the movements of both are the outcome of related projections of a common higher-dimensional ground. (Bohm 1980, p. 209)

This is a fairly unusual idea in the contemporary debate about the relation of matter and consciousness. It brings to mind Leibniz's idea of a pre-established harmony, but the difference is that Bohm emphasizes a constant correlated projection from a common ground, rather than a pre-established correlation. If you like, in Bohmian ontology the correlation is not pre-established, but rather established afresh at every moment. Yet Bohm is suggesting that there is kind of non-causal correlation between mind and body which is presumably in some ways similar to that which Leibniz had in mind.

Bohm then notes that, strictly speaking, it is too narrow to consider just the ground of mind and body, if one wants to understand what is happening in a given situation in which one typically finds oneself:

> Of course, even this ground of mind and body is limited. At the very least we have evidently to include matter beyond the body if we are to give an adequate account of what actually happens and this must eventually include other people, going on to society and mankind as a whole. In doing this, however,

> we will have to be careful not to slip back into regarding the various elements of any given total situation as having anything more than relative independence. (Bohm 1980, pp. 209–210)

The above point is important when trying to understand social phenomena. As human beings, we have relative independence from other human beings and from nature. Indeed, our freedom as individuals depends on such relative independence. Bohm, however, emphasizes the limits of such independence. Instead of thinking of human beings and nature as separate elements in causal interacion, we ought to see them, too, as correlated projections of a common ground:

> In a deeper and generally more suitable way of thinking, each of these elements is a projection, in a sub-totality of yet higher "dimension" ... it will be ultimately misleading and indeed wrong to suppose, for example, that each human being is an independent actuality who interacts with other human beings and with nature. Rather, all these are projections of a single totality. (Bohm 1980, p. 210)

This seems to imply a very radical kind of "social holism". It has important implications for our attempts to change social reality:

> As a human being takes part in the process of this totality, he is fundamentally changed in the very activity in which his aim is to change that reality which is the content of his consciousness. To fail to take this into account must inevitably lead one to serious and sustained confusion in all that one does. (Bohm 1980, p. 210)

Unfortunately Bohm is not very explicit in these sentences. At least I find it hard to understand more specifically what he means. Perhaps he means to say that if I try to change what I take to be reality, I will myself be fundamentally changed in this very activity. And if I do not take this into account, this will lead to confusion in all that I do. An example would have helped here to bring out what he had in mind. If he is correct in suggesting that human beings and nature are projections of a higher-dimensional reality, we should, of course, consider how this might affect our attempts to change reality. If reality and the human being are, strictly speaking, a unity, then changing reality means, at least to some degree, changing oneself. Perhaps this is what he had in mind.

In a similar way, Bohm thinks that when it comes to the side of mind, it is also necessary to consider a more inclusive ground:

> Thus, as we have seen, the easily accessible explicit content of consciousness is included within a much greater implicit (or implicate) background. This in turn evidently has to be contained in a yet greater background which may include not only neurophysiological processes at levels of which we are not generally conscious but also a yet greater background of unknown (and indeed ultimately unknowable) depths of inwardness that may be analogous to the "sea" of energy that fills the sensibly perceived "empty" space. (Bohm 1980, p. 210)

He makes an analogy between the "sea" of energy in empty space and the vast "unconscious" background of explicit consciousness. This question is crucially important for our understanding of the deeper nature of consciousness. Is consciousness merely a side-effect of the mechanical functioning of the brain, a fairly limited affair? Or is consciousness something much subtler, having depths of inwardness far beyond our ordinary consciousness? There is a tendency in contemporary cognitive neuroscience to opt for the former alternative. The trouble here, of course, is that whatever we assume the nature of consciousness to be may affect how we direct our attention and consequently what we will "see" the nature of consciousness to be in introspection (see Pylkkänen (2004b)). In order to avoid bias one way or the other, one thus has to approach the issue very carefully.

Bohm also connects his views with those of Freud:

> In some ways this idea of an "unconscious" background is similar to that of Freud. However, in Freud's point of view the unconscious has a fairly definite and limited kind of content and is thus not comparable to the immensity of the background that we are proposing. Perhaps Freud's "oceanic feeling" would be somewhat closer to the latter than would be his notion of the unconscious. (Bohm 1980, p. 217)

Bohm certainly was an advocate of the view that consciousness involves an "oceanic feeling", although he also thought we have a tendency to forget about it as a result of the way the "explicate order" dominates everyday experience. But surely various common experiences (e.g. aesthetic, religious, sexual, political), as well as perhaps experiences under the influence of alcohol and drugs illustrate what such an "oceanic feeling" refers to (cf. Pylkkö (1998)).

Bohm finally extends the analogy between the unconscious background of consciousness and empty space:

> Whatever may be the nature of these inward depths of consciousness, they are the very ground, both of the explicit content and of that content which is usually called implicit. Although this ground may not appear in ordinary consciousness, it may nevertheless be present in a certain way. Just as the vast "sea" of energy in space is present to our perception as a sense of emptiness or nothingness, so the vast "unconscious" background of explicit consciousness with all its implications is present in a similar way. That is to say, it may be sensed as an emptiness, a nothingness, within which the usual content of consciousness is only a vanishingly small set of facets. (Bohm 1980, p. 210)

This illustrates what we might call "Bohmian existentialism". According to the tradition of Western science, matter is the movement of atoms and fields in the void, and consciousness (if it is assumed to exist at all) is some kind of side-effect of such movement of atoms and fields. Bohm's view of matter suggests that ordinary matter is like a little ripple on the surface of the vast energy contained in empty space. Analogously, our usual content of consciousness exists on top of a vast unconscious background. This opens up the question whether human beings could sense or experience the unconscious background. Bohm certainly explored this dimension himself, especially in his many discussions with the Indian-born thinker Jiddu Krishnamurti (see, for example, Krishnamurti and Bohm (1999)).

3.8 Time in the Total Order of Matter and Consciousness

Bohm's new ontology seems to imply a new notion of time, and he next goes on to make more explicit this new notion. He first draws attention to the apparent difference between inner, psychological time and outer, physical time. Time as directly sensed in consciousness is variable (for example, a period may feel long or short) and relative to conditions; physical time seems absolute and independent of conditions. The theory of relativity changes all this. It implies that physical time is in fact relative, as it varies according to the speed of the observer. However, the reason why we do not typically realize this is that the "variation is significant only as we approach the speed of light and is quite negligible in the domain of ordinary experience" (Bohm 1980, p. 211).

Bohm next considers together quantum non-locality and the relativistic notion of space and time, which gives rise to some further interesting proposals:

> What is crucial in the present context is that, according to the theory of relativity, a sharp distinction between space and time can not be maintained (except as an approximation, valid at velocities small compared with that of light). Thus, since the quantum theory implies that elements separated in space are generally non-causally and non-locally related projections of a higher-dimensional reality, it follows that moments separated in time are also such projections of this reality. (Bohm 1980, p. 211)

I think what Bohm means here is that because space and time cannot be sharply distinguished, what appears as a non-local correlation between spatially separated elements at a given moment in a certain frame of reference may instead appear as a correlation between elements at two different moments in another frame of reference. Given that one was using the idea of correlated projections from the higher-dimensional reality to account for the non-local correlations in space, it seems natural to offer the same suggestion to account for the correlation between the moments. After all, one is examining the same correlated elements from different frames of reference. For Bohm, the implications are metaphysically very far-ranging:

> Evidently, this leads to a fundamentally new notion of the meaning of time. Both in common experience and in physics, time has generally been considered to be a primary, independent, and universally applicable order, perhaps the most fundamental one known to us. Now, we have been led to propose that it is secondary, and that like space... it is to be derived from a higher-dimensional ground, as a particular order. Indeed, one can further say that many such particular time orders can be derived for different sets of sequences of moments, corresponding to material systems that travel at different speeds. However, these are all dependent on a multidimensional reality that cannot be comprehended fully in terms of any time order, or set of such orders. (Bohm 1980, p. 211)

There are a number of very important points in the above quotation. The first is that time is not to be seen as fundamental but as derivative. It is a particular order that arises from the higher-dimensional ground. Secondly, there is an attempt to give an ontological interpretation to some of the more puzzling features of the theory of relativity. It is well known that the theory of relativity implies that time is not absolute but relative. But is this relativity at the level of reality or of appearances? Bohm suggests above a way of understanding how, say, clocks could actually move at different rates, depending on the speed of the observer. Basically, the higher the speed, the lower the rate of the clock (or of all the physical processes in that moving frame of reference, including the body of the observer, providing one way of explaining the twin paradox; see Bohm and Hiley (1985)). A given observer will experience a given sequence of moments at a given rate, suggesting a particular time order; another observer accelerating in relation to the former will experience a different sequence of moments, according to the predictions of the theory of relativity.

The third point in the above quote is that all time orders are dependent on a multidimensional reality that cannot be comprehended fully in terms of any time order, or set of such orders. This amounts to assuming that reality is infinite, both quantitatively and qualitatively. It also amounts to assuming that there is no fundamental level, of either explicate or implicate elements. One worry that arises with the Bohmian ontology is that the explicate order is nothing but a "shadow play" of the implicate order, thus denying the explicate order any real autonomy (contrary to what presumably was Bohm's intention when he says that the explicate order has "relative autonomy"). By assuming that there is no fundamental level, Bohm thinks he can avoid denying the reality of other levels. If there is no fundamental level, then there can be a sense in which each level is real and makes its contribution to reality (see Bohm (1986), p. 187).

Bohm next moves on to make some fairly exotic suggestions on the basis of the above view of time:

> Similarly . . . this multidimensional reality may project into many orders of sequences of moments in consciousness. Not only do we have in mind the relativity of psychological time . . . but also much more subtle implications. Thus, for example, people who know each other well may separate for a long time (as measured by the sequence of moments registered by the clock) and yet they are often able to "take up from where they left off" as if no time had passed. What we are proposing here is that sequences of moments that "skip" intervening spaces are just as allowable forms of time as those which seem continuous. [*Footnote:* This corresponds to the quantum theoretical requirement that

> electrons may go from one state in space to another, without passing through intermediate states.] (Bohm 1980, p. 211)

Time in the above view is understood as a sequence of moments. Bohm seems to imply that what is essential is the projection of such moments from the multidimensional ground, and not so much the fact whether the sequence is continuous or involves "skipping" intervening spaces. He seems to suggest that a given process (e.g. a discussion between two persons which extends over a period of years) can maintain its identity as a single process, even if there are discontinuities. Think of the root meaning of the word process, which is to proceed or to go step by step. Bohm suggests that the distances between the steps in a process can be fairly large, without this having any significant influence upon the process.

He then moves on to consider the relationship between the fundamental laws and time:

> The fundamental law, then, is that of the immense multidimensional ground; and the projections from this ground determine whatever time orders there may be. (Bohm 1980, p. 211)

This makes it clear that Bohm assumes time to be a secondary, derivative feature. Before time there is the multidimensional ground. In this ground a certain law prevails, and the outcome is projections of moments. Such projections form a sequence, and what we usually call time is then such a sequence of projected moments.

Now, when we think of causality in the everyday world, we may think of processes which follow the laws of classical physics; or else we may think of processes which for all practical purposes are unpredictable (e.g. a roulette wheel). How does Bohm reconcile his notion of the fundamental law with the more everyday notion of causal law and chance?

> Of course, this [fundamental] law may be such that in certain limiting cases the order of moments corresponds approximately to what would be determined by a simple causal law. Or, in a different limiting case, the order would be a complex one of a high degree which would approximate . . . a random order. These two alternatives cover what happens for the most part in the domain of ordinary experience as well as in that of classical physics. (Bohm 1980, pp. 211–212)

He then comes to consider a point that is very important for understanding the causal architecture of quantum domain, consciousness, and even biological phenomena:

in the quantum domain as well as in connection with consciousness and probably with the understanding of the deeper, more inward essence of life such approximations will prove inadequate. One must the go on to a consideration of time as a projection of multidimensional reality into a sequence of moments. (Bohm 1980, p. 212)

We have noted above that there seems to be a very deterministic quality to the implicate order. For example, the ink-in-fluid analogy, if interpreted as a model of the world, suggests a very deterministic world. It is as if God had placed all the droplets as she or he pleases, and all that is left is to unfold this implicate order. I have earlier said that Bohm thinks that reality is incomplete, and that there is the possibility of the creation of new potentialities, as well as new ways of unfoldment. But what could this mean more concretely?

As we have seen, he considers time to be a projection of multidimensional reality into a sequence of moments:

> Such a projection can be described as creative rather than mechanical, for creativity means just the inception of new content, which unfolds into a sequence of moments that is not completely derivable from what came earlier in this sequence or set of such sequences... movement is basically such a creative inception of new content as projected from the multidimensional ground. (Bohm 1980, p. 212)

In contrast,

> what is mechanical is a relatively autonomous sub-totality that can be abstracted from that which is basically a creative movement of unfoldment. (Bohm 1980, p. 212)

This makes it clear that Bohm thinks that creativity is a fundamental feature of the basic reality, while a mechanical, non-creative sequence is a special case. He considers the evolution of life in this context:

> ...the very word "evolution" (whose literal meaning is "unrolling") is too mechanistic in its connotation to serve properly in this context. Rather... we should say that various successive living forms unfold creatively and in the sense that later members are not completely derivable from what came earlier, through a process in which effect arises out of cause (though in some approximation such a causal process may explain certain limited aspects of the sequence). The law of this unfoldment

> cannot be properly understood without considering the immense multidimensional reality of which it is a projection (except in the rough approximation in which the implications of the quantum theory and of what is beyond this theory may be neglected). (Bohm 1980, p. 212)

By considering a concrete example of the evolution of life, Bohm makes it clear that he is not trying to explain the process of reality in terms of any simplistic notion of causal law or chance, or their combination. Instead, he suggests that something beyond these extremes is required. The key idea is that of "creative projection", the inception of new content into a sequence of moments. Such creativity in a given moment cannot be fully understood in terms of the previous moments. To understand its origin, one has to consider the multidimensional reality which is its ultimate source. Of course, there is a still a possibility of determinism in the Bohmian scheme in the sense that the projection of the new content is not arbitrary but is based on some potentiality in the implicate order of the multidimensional ground. Bohm discusses the possibility of creativity and its relationship with the question of determinism in his scheme in more detail in his article "Time, Implicate Order, and Pre-space" (Bohm (1986); see also Bohm and Peat (1987); Bohm and Biederman (1999)).

3.9 Metaphysics as a Proposal

We have now come to the end of going through Bohm's description of his overall implicate order approach. We have seen an attempt to achieve a grand synthesis in which questions of the nature of the cosmos, of matter in general, of life, and of consciousness have been brought together. The basic idea is that these can be considered projections of a common ground. Bohm calls this the "ground of all that is, at least insofar as this may be sensed and known by us, in our present phase of unfoldment of consciousness" (Bohm 1980, p. 212). But how can we know that there exists such a ground?

> Although we have no detailed perception or knowledge of this ground it is still in a certain sense enfolded in our consciousness, in the ways we have outlined, as well as perhaps in other ways that are yet to be discovered. (Bohm 1980, pp. 212–213)

The next question he considers is whether this ground is the absolute end of everything. Physicists, as is well known, often talk about the "theory of everything", a final and complete theory of the physical universe, in terms of the elementary particles and their interactions. Is Bohm trying to do basically the same sort of thing, albeit in a different way (for instance, by assuming the ground and projection to be fundamental, instead of the elementary particles and their interactions)? He answers:

> In our proposed views concerning the general nature of "the totality of all that is" we regard even this ground as a mere stage, in the sense that there could in principle be an infinity of further development beyond it. (Bohm 1980, p. 213)

There is a sense in which Bohm is doing metaphysics in the Aristotelian sense of studying "being qua (as) being". He is interested in understanding reality as it is in itself, not merely as it appears to us. For example, the whole notion of the ground is the result of an attempt to articulate what is implicit in our notions of consciousness and in our notions of quantum theory and cosmology. At the same time, he is also very modest and does not assume that his theories are true or complete. But how then are we to understand them?

> At any particular moment in this development [of our views] each such set of views that may arise will constitute at most a proposal. It is not to be taken as an assumption about what the final truth is supposed to be, and still less as a conclusion concerning the nature of such truth. (Bohm 1980, p. 213)

If the task of metaphysics (or at least of its general part, ontology) is to provide us with a view of the general nature of the totality of all that is, Bohm advocates great caution in this enterprise. The dream of the final truth is present not only in the hopes of some physicists; it is also implicit in the views of at least some of the contemporary philosophers engaged in metaphysics. In contrast, Bohm thinks that in metaphysics we ought to be satisfied with making proposals. But what does this mean? How should we understand a proposal about the general nature of the totality of all that is? Bohm's idea is that

> this proposal becomes itself an active factor in the totality of existence which includes ourselves as well as the objects of our thoughts and experimental investigations. (Bohm 1980, p. 213)

There is a strong tendency in philosophy to consider our beliefs and reality as separate domains. The theory of knowledge then focuses on questions like the truth of the beliefs and typically sees truth as either some kind of relation between beliefs and reality (e.g. correspondence), or between the beliefs themselves (e.g. coherence). Bohm emphasizes the fact that our beliefs are also some kind of existent entities or processes. They can thus be a part of the domain that they try to describe, and they may play an active role in giving shape to the reality via our actions. How should we then evaluate our beliefs?

> Any further proposals on this process, like those already made, have to be viable... one will require of them a general self-consistency as well as consistency in what flows from them in life as a whole. (Bohm 1980, p. 213)

The reference to "general self-consistency" resembles the coherence theory of truth which is widely discussed in contemporary epistemology (see, for example, Audi (1998)). The reference to "consistency in what flows from our proposals in life as a whole" reminds one of pragmatic theories of truth. It seems to me, however, that Bohm had a broader idea in mind than that typically considered in pragmatism. Indeed, this topic was central to the joint paper Bohm and I were working on in 1990–1991 (Bohm & Pylkkänen, unpublished). Bohm suggested there the term "overall coherence" to describe the aim of our proposals. His idea was that because reality is a process with many levels and orders in it, it is inevitable that incoherence constantly appears both in thought and in reality. However, what is possible is a movement towards coherence. Indeed, he suggested that we ought to consider cognition as essentially a "movement toward coherence". (I though this was such a succinct way of describing the point that I suggested that we use it as the title of our joint paper). "Overall coherence" did not merely mean the logical consistency of a set of beliefs (which is a sort of notion typically considered in the coherence theory of truth). It also referred to our psychological and physical state of being. We might still say that truth is a kind of coherence, but because Bohm thought it was overall coherence, he expressed this as saying that truth is part of what is. Thus truth is not just an abstract property, such as the correspondence between a set of beliefs and a state of affairs in objective reality. Nor is it just a relation of logical coherence that prevails between beliefs in a system. More strongly, truth is the overall coherence that can include a certain correspondence and logical consistency, but also includes the coherence of the state of being of the individual who is concerned with truth, as well as the environment. In this sense truth is "part of what is", a concrete reality we can experience, and are likely to lose as incoherence inevitably sets in at a subsequent moment. But once incoherence sets in, there is again the possibility of a new movement toward coherence (Bohm & Pylkkänen, unpublished). Remember that movement is a fundamental concept in Bohmian ontology. Strictly speaking, at the level of the ground of our consciousness and body, each one of us is a kind of movement. As we know too well, incoherence prevails very deeply in the human condition. Bohm did not deny the existence of this incoherence; on the contrary, as many of his publications show, he felt that we ought to give our urgent attention to it and go to its root. The optimistic side of his framework is exemplified in his thinking that movement toward coherence is possible, and is indeed the essence of cognition.

We have seen that Bohm encourages us to be aware of the fact that we are part of the totality of existence. Our views about the general nature of this totality are not merely passive representations or reflections of it, but may play some active role in the way the totality develops:

> Through the force of an even deeper more inward necessity in this totality, some new state of affairs may emerge in which both the world as we know it and our ideas about it may undergo an unending process of yet further change. (Bohm 1980, p. 213)

This underlines Bohm's view that the aim of metaphysics is not to seek a final view about the general architecture of existence. Of course, we could say that this view reflects his own assumption about that general architecture. He assumes that the totality of existence, including our knowledge, is movement; movement implies the possibility of change, and thus it is natural to expect that both reality and our knowledge of reality can change. In order to be self-consistent, he has to admit that his own view about the nature of reality as movement may turn out to have a limited range of applicability. All of this emphasizes that our metaphysical views ought to be offered as proposals, in the spirit of fallibilism: one day we may come to realize the limits of any proposal about the architecture of existence.

4

Active Information

4.1 Introduction

We have now gone through the main features of Bohm's implicate order scheme, as it applies to matter – both inanimate and animate – and consciousness. We have also examined what it has to say about the general architecture of existence, for example about such basic issues as the nature of causality and the nature of time. Finally, we have considered how the implicate order theory should be taken or understood. According to Bohm, much harm in science, and even more broadly in society, is caused by our taking our general world views as final and absolute truths. Thus, when launching the implicate order scheme, he was keen to emphasize that it is merely a proposal. If you like, it is a tool that can be used to see. One thinks here about the root meaning of the word "theory". "Theory" derives from the Greek "theoria", which has the same root as "theatre", namely "theaomai", which means "to view". In this spirit let us by all means "enjoy the show", the new view that a new theory can provide; but let us not forget that our theory, like all views, is likely to be incomplete. Theories are not the sorts of things that provide us with absolute and final truths. Likewise, a theory about the general architecture of existence is not likely to provide us with a complete and final view of this architecture.

However, I have suggested from the beginning of this study that a change in our general world view can have all sorts of significant implications for our more specific studies. A new notion of matter might well have implications for, say, studies which try to understand the relationship between mind and matter. A new understanding of the nature of causality and time can have implications for a wide range of fields. I thus think that Bohm's implicate order scheme, with its new notions of matter, causality, and time, opens up many possibilities.

In this book our focus is on the possibilities Bohm's thinking opens up for the understanding of mind, matter, and their relationship. We have already seen a number of these in the previous chapters. In the light of the implicate order scheme, the main reason why it has been so difficult to relate matter and consciousness is that the traditional scientific world view gives such primary importance to the explicate order, while as Descartes observed, consciousness has properties that seem to go beyond existence in an explicate order. Bohm's general line of approach to matter and consciousness is to suggest that the fundamental order of both is the implicate order, while the explicate order is a special, derivative case of the implicate order. This then leads to the idea that there exists an underlying ground, of which mind and body are aspects or correlated projections.

Bohm published these ideas around 1980, most notably in the final chapter of *Wholeness and the Implicate Order*, which we went through in Chaps. 2 and 3. However, it seems he was not fully satisfied with his discussion of the relationship between mind and matter in the implicate order scheme. Indeed, he admitted that the implicate order, in the form proposed in his 1980 book, was at a relatively early stage of development, both as a theory of matter and as a framework within which to discuss the relationship between mind and matter. It should be seen as

> *a general framework of thought* within which we may reasonably hope to develop a more detailed content that would make progress toward removing the gulf between mind and matter. (Bohm 1990, pp. 273–274).

He notes that even on the physical side, i.e. as a theory of physics, the implicate order

> lacks a well-defined set of general principles that would determine how the potentialities enfolded in the implicate order are actualized as relatively stable and independent forms in the explicate order. (Bohm 1990, p. 274).

What about the hologram and the ink-drop analogies? – Do these not provide us with such principles? The trouble with these is that both *presuppose* relatively stable and independent forms in the explicate order. The hologram stores information about pre-existing objects, and the ink-drop analogy involves enfolding a pre-existing droplet. But where do these kinds of objects come from in the first place? It is, of course, one of the most important challenges of the implicate order programme to be able to answer this question. The analogies of the hologram and the ink-drop, however, cannot as such provide the answer. This points to the need to extend the implicate order framework before it can be considered a satisfactory theory for physics.

Bohm further notes that a similar set of principles is also absent on the mental side (Bohm 1990, p. 274). He is not very explicit about what he means by this. Now, he suggested that the same basic architecture that applies for matter also applies for mind. Thus, we can see in a number of mental phenomena that the implicate order prevails; but the explicate order, too, plays an important role in conscious experience (e.g. as the spatio-temporally organized "virtual reality" or inner show of sense experience) and in mental phenomena more generally (e.g. in being the ground of memory).

Presumably Bohm is above acknowledging that he has not provided a specific theory about how the "virtual reality" or the explicate order of conscious experience arises from the underlying, more fundamental implicate order of the mind. In the Bohmian scheme such a theory might, for example, show how the "virtual reality" of conscious experience (with its spatio-temporal organization) arises from the underlying "pre-space" of the enfolded information in the nerve cells. This could be done analogously to how Bohm and Hiley have tried to show how the physical space can arise from the underlying pre-space (see Bohm and Hiley (1984)). Indeed, Bohm (1986) suggested that the virtual reality of consciousness is likely to arise in this way. But we need to specify the principles which govern the way the virtual reality of the explicate order of consciousness arises from the implicate order.

Bohm also admits that the implicate order theory does not provide a clear idea of *just how* the mental and material sides are to be related (Bohm 1990, p. 274). Presumably, he was not fully satisfied with the idea of matter and consciousness as correlated projections, or the general idea that the relationship between matter and consciousness is that of "mutual enfoldment". Something more needs to be said. Therefore, he suggested that if one wants to tackle the mind–matter problem more coherently, it is likewise necessary to extend the framework of the implicate order.

Bohm first attempted such an extension by introducing a general theoretical notion he called "soma-significance", which we touched on briefly in Chap. 1 (Bohm 1985). His second and more detailed attempt is based upon the ontological interpretation of the quantum theory, which we have also briefly discussed. In this chapter we shall consider in more detail the ontological interpretation and how it extends the implicate order scheme, both as a theory of physics and as a theory of the mind–matter relationship.

4.2 The Ontological Interpretation of the Quantum Theory

As we have seen in Chap. 1, the ontological interpretation is based on an interpretation of quantum theory in terms of "hidden variables", which Bohm originally proposed in 1952, and later developed especially in co-operation with his long-time colleague Basil Hiley (Bohm & Hiley 1993). Bohm presented in his 1952 papers both a "particle theory", i.e. an ontological model of quantum particles (such as an electron), and, in an appendix, also a "field theory", i.e. an ontological model of how the electromagnetic field of classical physics is "quantized" to give rise, at least momentarily, to "bullets of light" (i.e., quanta of energy, or photons), as quantum theory famously says it does.

Now, remember that Bohm felt that the ontological interpretation can do two things to make the implicate order more specific: first, to show how the explicate order arises out of the implicate order in physical processes, and second, to provide a more specific idea about how mind and matter are related. To see to how the explicate order arises out of the implicate order in physical processes, it is useful to consider the field theory, i.e. the ontological interpretation of the electromagnetic field. To obtain a more specific idea about how mind and matter are related, it is useful to consider both the particle theory and the field theory. Let us start with the particle theory, as this also makes it easier to later understand the field theory.

As we saw in Chap. 1, the starting point of the ontological interpretation of quantum particles (e.g. electrons) is simply to assume that

> the electron, for example, is a particle which follows a well defined trajectory (like a planet around the sun). But it is always accompanied by a new kind of quantum field. (Bohm 1990, p. 276)

This assumption may sound innocent, but it differs in some ways radically from the way in which one thinks of an electron in the usual interpretation of quantum theory. In the usual interpretation, the fundamental laws of the quantum theory are expressed with the aid of a wave function. The idea is that the wave function is the most complete possible specification of the physical state of an individual system such as an electron. This would seem to suggest that the electron is a wave. However, when we observe the electron in an experiment, we always find it as a little particle. What, then, is the relationship between the description of the electron as a wave before the measurement and the actual observation of the electron, which reveals it as a little particle? We mentioned in Chap. 1 some different ways of thinking about this.

There is a "minimalist interpretation" which assumes that the wave function should not be taken as a description of the electron, but rather as a mathematical tool which allows us to calculate the probability finding the electron, conceived of as a little particle, in a given small region of space.

Then there is the "collapse interpretation", which assumes that the electron is a wave before measurement, a wave, however, which collapses into a small region in a measurement, giving rise to the small particle-like manifestation we actually observe.

We also briefly mentioned yet another interpretation according to which the electron is a wave before measurement. When we carry out the measurement, we find the electron as a particle in a small region of space. However, the wave has not collapsed, but its "disappearance" is explained by assuming that the universe branched in the measurement. Those parts of the wave function that we are not observing are thought to exist in parallel universes that came into being as a result of the branching. In one of these universes there is a replica of me seeing a particle at location A, in another universe there is another replica of me seeing a particle at location B, and so on, covering all locations where the square of the absolute value of the wave function is non-zero. This is the famous "many-worlds" interpretation.

Finally, Bohm's ontological interpretation assumes that the electron is both a particle and a wave before measurement. In the measurement we see the particle aspect. The wave aspect guides the particle aspect by giving rise to a new potential, the quantum potential. As in all the other interpretations, the wave function (which describes the wave aspect) can be used to calculate the probabilities of finding the particle aspect of the electron in a given region of space.

The meaning of the ontological interpretation can be brought out further by considering in more detail how it differs from the others. It agrees with the minimalist interpretation in assuming that the wave function can be used to calculate probabilities of finding the electron as a little particle in a small region. However, the ontological interpretation says more than this. Contrary to the minimalist interpretation it assumes that the wave function describes a field that exists "out there", in some ways similar to the electromagnetic field, for example. Further, unlike the minimalist interpretation, it assumes that the electron has a particle aspect all the time, and in particular that the particle aspect is there prior to measurement. The minimalistic interpretation is agnostic about whether or not the electron exists as a particle before measurement, although it admits that we find the electron as a particle in a measurement.

The ontological interpretation agrees with the collapse interpretation that the wave function should be seen as a description of objective reality, suggesting that quantum systems involve the objective existence of a new type of field. However, the ontological interpretation does not assume that the wave function provides a complete description of the system before measurement. There exists something more, namely the particle aspect of the electron. But this means that there is no need to assume the collapse of the wave function to explain why we see a particle in every measurement. What we see is the particle aspect of the electron which was there all along, guided in its motion by the wave aspect.

Finally, the ontological interpretation agrees with the many worlds interpretation that the wave function should be seen as a description of objective reality, again suggesting that quantum systems involve the objective existence of a new type of field. It also agrees with the latter in that no collapse of the wave function is postulated. However, it differs from the latter in assuming that the electron also has a particle aspect all the time. Thus, there is no need to postulate that the universe branches in order to explain why we see the particle in every measurement. The reason we see the particle is simply that we assumed from the very beginning that there exists a particle.

We can further try to understand the ontological interpretation by examining how it accounts for the well-known two-slit experiment. Let us begin this by stating the problem in a standard way, following Feynman (1965) and Pagels (1982). In Fig. 4.1 the two upper pictures remind us how particles and waves behave according to classical physics when they go through a partition which contains two adjacent slits, one or both of which may be open. The first picture illustrates, using bullets as an example, how particles behave in the situation. P_1 describes the distribution of bullets in a wall when only slit 1 is open, P_2 describes it when only slit 2 is open and P describes the distribution when both slits are open. It is obvious that the opening of the second slit increases the possible places at which bullets may arrive. P is simply the sum of P_1 and P_2.

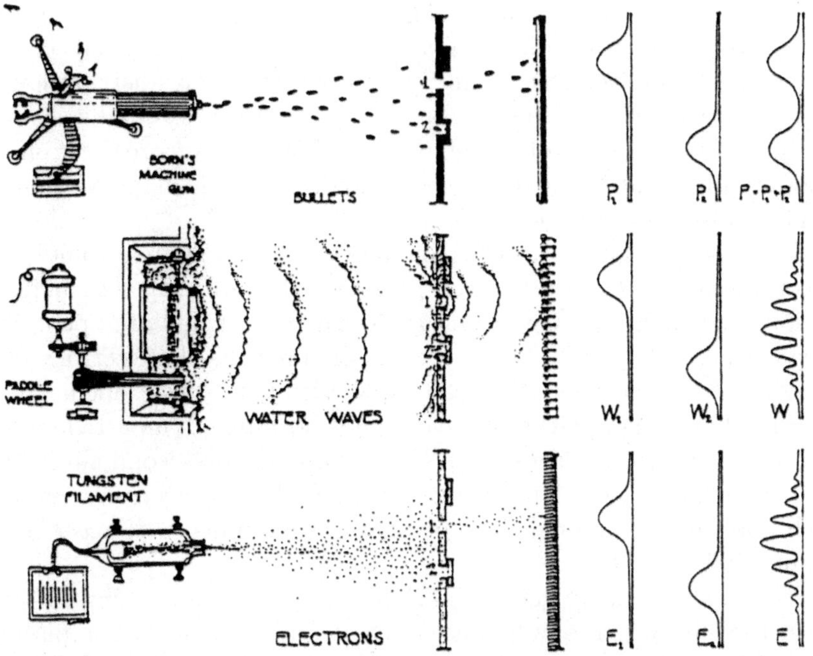

Fig. 4.1. The behavior of bullets, water waves, and electrons in the two-slit experiment. It is easy to understand why bullets and water waves arrive at the places they do in the detector placed behind the slits. It is very strange, however, that the electrons should form an interference distribution (E) when two slits are open. (The figure is from Pagels (1982).)

In the second picture, water waves illustrate how classical waves behave in a corresponding situation. W_1 describes the intensity distribution of waves on a detector when just one slit is open. This resembles P_1, which we got with bullets in the corresponding situation. The situation is the same with W_2 and P_2. If we now open both slits, we get an intensity distribution W which is different from P, for W is not a sum of W_1 and W_2. According to W, strong waves arrive at certain regions, whereas in between these regions no waves arrive at all. The reason for this is that in some places the waves coming from slits 1 and 2 strengthen each other (they interfere constructively when they are in phase), whereas in other places they cancel each other out (they interfere destructively when they are out of phase). In the two-slit experiment we get interference fringes with classical waves.

What happens with quantum systems? In the lowest picture, electrons "boil" out of a hot filament, and they move through the slits toward a photographic plate. The picture reveals that the electrons have astonishing wave-like properties. The intensity distribution E of an ensemble of electrons is similar to the water wave distribution W in the corresponding situation. In this sense the electrons have wave-like properties. On the other hand, individual electrons register on the photographic plate at a single spot, one by one, which shows that an electron also has particle-like properties.

If we now close slit 2, electrons arrive at the photographic plate according to E_1. The key point here is that electrons now can arrive at regions at which they never can arrive when both slits are open. This means that the opening of the second slit prevents the electrons from arriving at regions where they can arrive with just one slit open. This is bizarre, if one assumes that the electron is nothing but a particle. For if electrons are just particles, then the opening of the second slit should in general be expected to *increase* the number of electrons reaching each point in the photographic plate, and should never *decrease* it (cf. Bohm (1957, p. 72)).

Notice also that we get an interference pattern even if the electrons enter the slit system one by one. Each electron arrives at the plate as a particle, giving rise to a little spot, but gradually these spots build up an interference pattern characteristic of waves. The interference pattern is thus not the result of a mutual interaction between the electrons. Rather, the opening of a second slit can prevent a *separate* and *independent* electron from reaching certain points in the plate that it could reach if that slit were closed (cf. Bohm (1957, pp. 72–73)). These points are those where waves in this situation would interfere destructively or cancel each other out. An individual electron avoids these points, thus strongly suggesting that it has wave-like properties. The assumption that individual electrons have wave properties then explains why a set of such electrons can give rise to an interference pattern. On the other hand, if we assume that the electron only has particle properties, it seems impossible to explain why it obeys the mathematics of wave motion in the peculiar way shown by the two-slit experiment.

Bohm describes the above as follows:

> In the usual interpretation of the quantum theory, the origin of this interference pattern is very difficult to understand. For there may be certain points where the wave function is zero when both slits are open, but not zero when only one slit is open. How can the opening of a second slit prevent the electron from reaching certain points that it could reach if this slit is closed? If the electron acted completely like a classical particle,

> this phenomenon could not be explained at all. Clearly, then, the wave aspects of the electron must have something to do with the production of the interference pattern. Yet, the electron cannot be identical with its associated wave, because the latter spreads out over a wide region. On the other hand, when the electron's position is measured, it always appears at the detector as if it were a localized particle. (Bohm 1952, p. 173)

Bohm neatly brings out the basic challenge presented by the two-slit experiment. On the one hand, if we assume that the electron is just a particle (as classical physics does), we cannot explain why the opening of the second slit prevents it from reaching certain points. On the other hand, if we assume that the electron is just a wave, we cannot explain why we always find it as a localized particle in a measurement. So what is the electron, then? It seems it cannot be just a particle, and it cannot be just a wave evolving according to Schrödinger's equation (which in this situation predicts that the wave is spreading out).

One might think that the reasonable thing to do would be to observe the movement of the electron. However, in the case of the two-slit experiment (and at the quantum level more generally) we do not, with the type of experiments that can be carried out today, directly observe the electron moving, either as a particle or as a wave. Attempts to observe the movement of the electron tend to destroy the interference pattern. This is reflected in Heisenberg's famous indeterminacy principle. We thus have to carefully think about the meaning of the results that we can observe in current experiments. What we observe in the two-slit experiment are spots appearing in a photographic plate, one by one, gradually forming an interference pattern. Because the electron appears as a spot, it is reasonable that we attribute a particle property to it. And because the electrons collectively produce a wave pattern, it is reasonable that we attribute a wave property to each individual electron.

As we cannot observe the electron in its movement, we need to make an "educated guess" about what it might be like when it is not observed, should we abandon positivist strictures and be interested in making such guesses (for example, because in the history of science it has frequently been fruitful to make such guesses). The evidence suggests that any such guess has to attribute the electron both a wave property and a particle property. Or more precisely, the electron has to be an entity or a process which is capable of manifesting both wave properties and particle properties.

One, it would seem very natural, way to think about this is to follow Bohm in assuming that the electron is a particle, always accompanied by a new type of quantum field. As we will soon see, in this way we can tell a coherent story about what happens in, say, the two-slit experiment. Yet it is important to remember at the outset that to assume that the electron is a particle is not as innocent as it may sound. In Chap. 1 we referred to the relativistic difficulties with thinking about the electron as a particle (whether point-like or extended). Bohm and Hiley themselves suggest that the idea of a particle is no longer relevant at the temporal level of the Planck time 10^{-43} s, where it has to be replaced by the notion of a succession of waves, in the spirit of the implicate order. But if Bohm and Hiley admit that the notion of the electron as a particle is limited, why do they then spend so much time (e.g. hundreds of pages in their 1993 book) discussing it? Remember that Bohm and Hiley assume that theories are tools – ways of looking which can provide a coherent view in some domain, but can fail beyond this domain. Taken in this way, the ontological interpretation of the "particle theory", with its notion of electron as a particle accompanied by a field, can be very useful in a certain domain, even if it may have to be dropped in a more fundamental theory. It is important to keep this in mind as we discuss the particle theory in this book.

Let us now see how the particle theory accounts for the two-slit experiment. Consider an electron coming from the filament towards the slits and the screen. Its quantum wave generally precedes it. The wave goes through both slits and produces an interference pattern after it has passed the slits. Each electron follows a well-defined path, going through one slit or the other. It is now acted upon not only by the classical potential V but also the quantum potential Q. The quantum potential is derived from Schrödinger's equation and is expressed mathematically as

$$Q = \frac{-\hbar^2}{2m}\frac{\nabla^2 R}{R} \tag{4.1}$$

What is important here is the way the quantum potential Q depends upon R, which latter is the amplitude of the quantum field (for $\hbar$ is Planck's constant h divided by 2π and m is the mass of the particle). Note first that R appears in both the denominator and the numerator. This is important because it means that R can be multiplied by an arbitrary constant without changing the quantum potential. Size doesn't matter here, as a small wave can have exactly the same effect as a large one! What does matter, then? Note that Q depends on the second spatial derivative of R. This reflects the way in which R changes over space, or the form of the wave. Thus, Q depends only upon the form of the quantum wave.

Before the slits, R is constant, its second spatial derivative is zero, and thus Q is also zero. After the quantum wave passes through the slits, R is, however, no longer constant but can vary in a complicated way as the two waves spreading out from the two slits meet and interfere. Thus, after the particle passes through the slit system, it will encounter a quantum potential which changes rapidly with position. As a result of the work of Chris Philippidis, Chris Dewdney, and Basil Hiley (1979) it is actually possible to visualize this potential (Fig. 4.2).

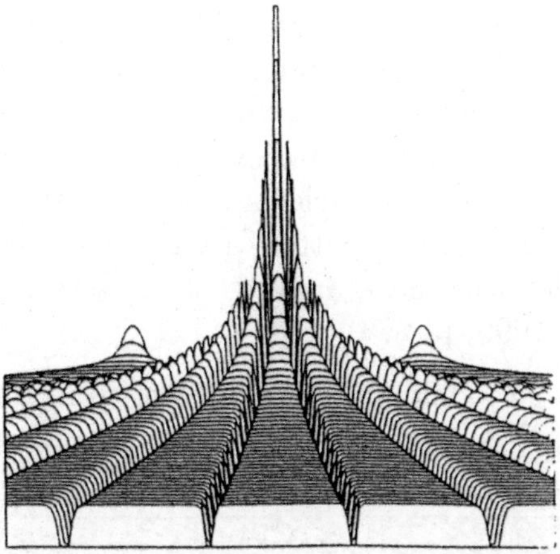

Fig. 4.2. The quantum potential for the two-slit interference experiment

We are looking at the potential as if we were at the photographic plate (so that the electrons are coming through the slits towards us). As Bohm describes it,

> the quantum potential consists of a series of "plateaus" separated by deep "valleys". When an electron crosses these "valleys", it is sharply accelerated. So the electrons are deflected even in the empty space in front of the slits by the quantum potential, and this deflection may still be large even far from the slits. (Bohm 1990, p. 277)

If one of the slits is now closed,

> [the quantum potential] is correspondingly altered, because the ψ-field is changed, and the particle may then be able to reach

certain points which it was unable to reach when both slits were open. (Bohm 1952, p. 174)

Let us look in more detail at how Bohm explains that we get the correct intensity distribution in this model:

> ... in a typical experiment of this kind, the source of electrons is a hot filament, behind the slits, out of which they may be thought of as "boiling" with a random statistical variation of initial positions (i.e., appearing here and there by chance). Each electron follows a particular path, going through one slit or the other, and it arrives at the detecting screen as an individual particle, producing, for example, an individual spot in a photographic plate located at the screen. In its movement, the electron is affected by the quantum potential, which ... is determined by the wave that in general precedes the particle. However, if we follow the whole set of trajectories, which represents an initially random distribution of particles ... these are "bunched" systematically into a fringe-like pattern (which will become apparent after many electrons have arrived at the screen in front of the slits). (Bohm 1990, p. 278)

This can be seen by looking at the trajectories in Fig. 4.3.

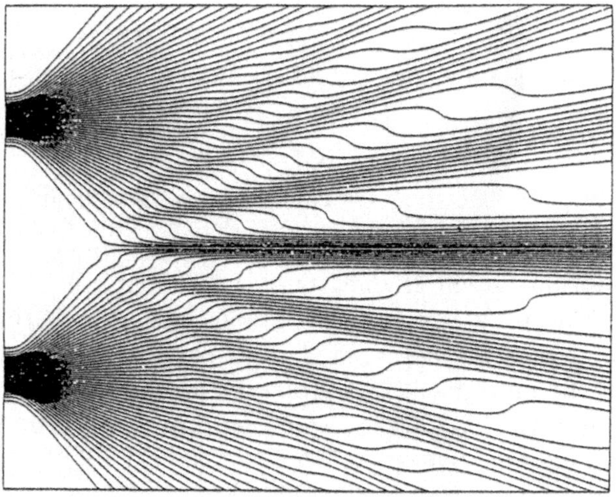

Fig. 4.3. Particle trajectories for the two-slit interference experiment

Bohm's model thus appears to provide an *intelligible explanation* of perhaps the greatest mystery of the two-slit experiment, namely how does the electron "know" whether the second slit is open or not. It "knows" this because of what its quantum field "tells" it. In this way it is possible for the electron to be a particle, and yet manifest wave-like properties statistically.

Bohm emphasizes that it is essential for this explanation that the quantum potential depends only on the form of the wave, so that it can be strong even when the wave intensity is weak. This explains the typically quantum mechanical, non-local feature, namely that even distant parts of the environment (for example, opening and closing of slits which could, in principle, be millions of miles away) can affect the motion of the particle in a significant way. This is contrary to Newtonian physics, in which the magnitude of a force diminishes as distance increases.

It is here that Bohm introduces a notion which, as we shall see, also becomes important for his mind–matter theory:

> I would like to suggest that we can obtain a further understanding of this feature [of non-locality] by proposing a new notion of active information that plays a key role in this context. The word in-form is here taken in its literal meaning; i.e., to put form into (rather than in its technical meaning in information theory as negentropy). One may think of the electron as moving under its own energy. The quantum potential then acts to put form into its motion, and this form is related to the form of the wave from which the quantum potential is derived. (Bohm 1990, p. 279)

Bohm then goes on to point out that there are many analogies to the notion of active information in our general experience:

> Thus, consider a ship on automatic pilot guided by radar waves. The ship is not pushed and pulled mechanically by these waves. Rather, the form of the waves is picked up, and with the aid of the whole system, this gives a corresponding shape and form to the movement of the ship under its own power. Similarly, the form of radio waves as broadcast from a station can carry the form of music or speech. The energy of the sound that we hear comes from the relatively unformed energy in the power plug, but its form comes from the activity of the form of the radio wave; a similar process occurs with a computer which is guiding machinery. The "in-formation" is in the programme,

> but its activity gives shape and form to the movement of the machinery. Likewise, in a living cell, current theories say that the form of the DNA molecule acts to give shape and form to the synthesis of proteins (by being transferred to molecules of RNA). (Bohm 1990, p. 279)

Bohm's basic proposal, then, is to *extend* this more ordinary notion of active information to matter at the quantum level. The information at the quantum level is *potentially active* everywhere, but *actually active* only where this information enters into the activity of the particle – just as the radio wave is active where the receiver is. He notes that such an extension of active information to the quantum level implies the prima facie strange idea that the electron may have an inner structure, but he goes on to argue that there is a possibility for such a structure between 10^{-16} cm and 10^{-33} cm.[1]

More of the new features of the quantum potential are revealed when one examines the many-body system. First of all, the quantum potential implies a non-local interaction between particles. Secondly, the quantum potential

> depends on the "quantum state" of the whole system in a way that cannot be defined simply as a pre-assigned interaction between all the particles. ... the wholeness of the entire system has a meaning going beyond anything that can be specified solely in terms of the actual spatial relationship of the particles. This is indeed the feature which makes quantum theory go beyond mechanism of any kind [and it is] the most fundamentally new ontological feature implied by the quantum theory. (Bohm & Hiley 1987, p. 331)

Bohm and Hiley also say that the quantum potential

> describes a part of the change with time of the active information that depends in a non-local way and in detail upon the whole wave function (i.e., what is commonly called the quantum state). (Bohm & Hiley 1993, p. 62)

Another important feature which comes out when considering the many-body system is that although the quantum field for a single par-

[1] The work of Hans Dehmelt (1990) seems relevant here. He has estimated that the electron radius is 10^{-20} cm – that is, one hundred billion billion electrons could fit within 1 cm. The electron is tiny indeed, but the key point is that it does not seem to be a point particle and may even have a composite structure. At any rate, this increases the plausibility of Bohm's idea that the electron can respond to the quantum wave in a subtle way.

ticle can be thought of as existing in ordinary (3+1)-dimensional space, the quantum field of the many-body system is in configuration space, which has ($3N$+1) dimensions, where N is the number of particles. How is it possible to give a physical interpretation of a multi-dimensional field? This question has traditionally been considered one of the most serious criticisms of realistically interpreted "guide-wave" models (e.g. Putnam (1975, p. 134); Bohm in Weber (1986, p. 111); Bohm and Hiley (1987, p. 332)). In this context Bohm and Hiley write:

> The fact that the wave function is in configuration space clearly prevents us from regarding the quantum field as one that carries energy and momentum that was simply transferred to the particles with which it interacted (thus effectively pushing or pulling mechanically on the latter). This is a further factor in addition to the form dependence of the activity of the field which leads us to consider the interpretation of this field as active information. The multidimensional nature of this field need not then be so mysterious, since information can be organized into as many sets of dimensions as may be needed. (Bohm & Hiley 1993, p. 61)

To give an analogy, even a computer could organize information into an abstract ($3N + 1$)-dimensional configuration space, by forming a suitable table of numbers, dimensionality being the number of independent coordinates in the table (Bohm and Pylkkänen, unpublished).

The wave function is thus defined in the configuration space of all the particles, and thus information at the quantum level is generally ordered in a multidimensional space rather than in the ordinary space of three dimensions. Bohm and Hiley point out that

> [t]he fact that the wave function is in configuration space implies that we have to look more carefully into the meaning of active information in such a context. First of all we may consider its implications for all the motions of the particles. These now respond in a correlated way to what is, in effect, a *common pool of information.* This information guides the particles according to the condition $\boldsymbol{p}_i = \nabla_i S(\boldsymbol{r}_1, \ldots, \boldsymbol{r}_N, t)/m$, which, of course, leads to a generally non-local quantum potential... (Bohm & Hiley 1993, p. 60; emphasis added; p = momentum, S = phase.)

This common pool of information reveals its activity in superconductivity:

> ...at ordinary temperatures, electrons moving inside a metal are scattered in a random way by various obstacles and irregularities in the metal. As a result, there is a resistance to the

flow of electric current. At low temperatures, however, the electrons move together in an organized way, and can therefore go around such obstacles and irregularities to re-form their pattern of orderly movement together [see Fig. 4.4]. Thus they are not scattered, and therefore the current can flow indefinitely without resistance. (Bohm 1990, p. 280)

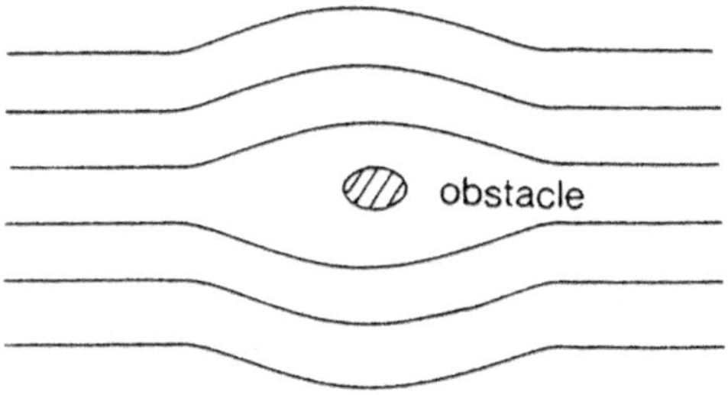

Fig. 4.4. Superconducting current flowing around an obstacle

Bohm then uses the fact that the quantum potential for the whole system constitutes a non-local connection to explain how the organized and orderly pattern of electrons moving together without scattering is possible. He even makes an analogy to a ballet dance, in which all the dancers, guided by a common pool of information in the form of a score, are able to move together in an organized and orderly way, to go around an obstacle and re-form their pattern of movement (Bohm 1990, p. 281).

Another key feature of the ontological interpretation is that it provides an elegant way to think about a notoriously difficult issue in quantum mechanics, namely the relationship between the quantum level and the classical level. We might think that surely these theories apply to the same reality. We might further think that quantum theory is the more general and accurate theory, and classical physics can be obtained as a special case in some suitable limit, a special case which provides a good approximate description of some subset of the physical world, the so-called classical domain. As I mentioned in Chap. 1, things were not that simple in the usual interpretation of quantum theory. Bohm made it very clear in his 1951 textbook *Quantum Theory* that the usual interpretation presupposes the existence of the classical level, and in particular a measurement apparatus that has a classically describable part. In contrast, the ontological interpretation presupposes such a quantum ontology from which the classical world can be derived whenever the quantum potential is negligibly small. Bohm and Hiley present their view of how the classical world emerges from quantum ontology in Chap. 8 of their *Undivided Universe.*

We have thus seen that Bohm and Hiley's quantum ontology contains several new features. The quantum potential depends only upon the form of the field, not upon its intensity, and its operation can thus be best understood in terms of the notion of active information, which is new in physics. The form dependence of the quantum potential makes possible a wholeness of the system and the distant features of its environment, something Niels Bohr also emphasized, though in a different way.

The quantum potential also explains how there can be a non-local correlation between systems, the radical feature of quantum theory which we discussed in Chap. 2. According to Bohm and Hiley, the most fundamentally new aspect of the quantum theory, brought to focus by the ontological interpretation, is that the quantum potential depends on the "quantum state" of the whole system in a way that cannot be defined simply as a pre-assigned interaction between all the particles. Further, the multidimensionality of the many-body wave function was seen as an additional reason for interpreting the field in terms of active information. A many-body system, whose wave function has not factorized into independent parts, can be described by saying that there is a common pool of information to which each individual particle responds.

Finally, it is possible to show that the effect of the quantum potential is generally negligible at the large-scale level. In this way we can understand why quantum effects do not generally affect the movement of macroscopic bodies, and classical physics provides a good approximate description. Note, however, that the ontological interpretation allows that the quantum potential need not always be negligible in macroscopic conditions. Thus it provides an elegant way to discuss such macroscopic quantum effects as superconductivity and superfluidity.

We have now gone briefly over the main features of the ontological interpretation of the particle theory. Let us now consider the ontological interpretation of the field theory. After that, we will consider first how Bohm felt the ontological interpretation can be used to extend the implicate order as a physical theory. Afterwards, we shall see how the ideas contained in the ontological interpretation of both the particle and field theory can be used to extend the implicate order framework in such a way as to enable a better notion of how the mind and matter are related.

4.3 The Ontological Interpretation of Quantum Field Theory

I mentioned above that Bohm applied the ontological interpretation in his 1952 paper not only to quantum particles, such as electrons, but also to the electromagnetic field. As is well known, the quantum theory originated when Planck and Einstein studied certain experiments in which matter exchanged energy with light. Their conclusion was that

> the light transmits energy to matter in the form of "quanta" or bundles, of size $E = hf$, where f is the frequency of the light wave and h is a universal constant, which was later called Planck's constant (Bohm 1957, p. 70)

According to classical physics, light consists of electromagnetic waves. To do justice to the experimental facts of the quantum domain, the electromagnetic field thus has to be "quantized", giving rise to "quantum field theory". The ontological interpretation provides one way of doing this, and Bohm did it for the electromagnetic field in his 1952 papers.

Twentieth-century developments in physics (e.g. the discovery of antimatter) suggest that, strictly speaking, all matter and energy, including "particles" such electrons, protons, and neutrons, ought to be understood in terms of quantum fields, and thus quantum field theory should be applied to them. This suggests that Bohm's ontological interpretation of the "particle theory", which we discussed in the previous section, cannot be a fundamental theory, although it may be very useful in some domain. Indeed, when sketching his more general view of the physical universe in a 1987 article "Hidden Variables and the Implicate Order", Bohm himself was keen to drop the particle theory, which he saw as "the mechanical starting point" (Bohm 1987a, p. 42), and opt for quantum field theory even for "fermionic" particles like electrons.[2] However, one should note that it has not been straightforward to develop a quantum field theory for particles like electrons in

[2] The "elementary particles" of physics are divided into two categories, depending on the nature of their spin, which in turn determines which statistics they obey in their behaviour. Particles of half-integral spin (which include the particles making up "ordinary matter" such as electrons, protons, and neutrons) obey the so-called Fermi–Dirac statistics and are called fermions. It is characteristic of fermions that they can never enter the same quantum state. Particles of integral spin obey the so-called Bose–Einstein statistics and are thus called bosons. There are massless bosons (e.g. the photon, or the particle-like manifestation of the electromagnetic field) and bosons with mass. Bosons can enter into the same quantum state, giving rise to the possibility of a macroscopic quantum effect, the Bose–Einstein

Bohm's ontological interpretation scheme. Bohm himself, for one thing, did not succeed in doing this. John Bell made some proposals to this effect in 1987, and there have recently been a number of similar efforts (see especially Nicolic (2005)).

Let us now move on to consider quantum field theory, and the way it looks when one applies Bohm's ontological interpretation scheme to it. Bohm first describes the way one usually accomplishes a quantum mechanical field theory. One starts with

> the classical notion of a continuous field (e.g. the electromagnetic) that is spread out through all space. One then applies the rules of the quantum theory to this field. The result is that the field will have discrete quantized values for certain properties, such as energy, momentum, and angular momentum. Such a field will act in many ways like a collection of particles, while at the same time it still has wave-like manifestations such as interference, diffraction, etc. (Bohm 1987a, p. 42)

Bohm points out that in the usual interpretation of quantum field theory, there is no way to understand how this comes about:

> One can only use the mathematical formalism to calculate statistically the distribution of phenomena through which such a field reveals itself in our observations and experiments. (Bohm 1987a, p. 42)

In other words, the same sort of minimal interpretation that has ruled quantum particle theory has also dominated quantum field theory, making it very difficult to understand the latter intuitively. The ontological interpretation can arguably provide a clear hypothetical picture of what happens in a typical experiment with quantum particles. Can it do the same with quantum fields? Here is how Bohm describes the extension of the ontological interpretation to the quantum field theory:

> ... the actuality will be the entire field over the whole universe. Classically, this is determined as a continuous solution of some kind of field equation (e.g. Maxwell's equations for the electromagnetic field). But when we extend the notion of the... [ontological] interpretation to the field theory, we find that these equations are modified by what I called the super-quantum potential. This is related to the activity of the entire field as the

---

condensate. This was predicted in the early 20th century by Bose and Einstein, and first created and observed in the University of Colorado in the 1990s.

> original quantum potential was to that of the particles. As a result the field equations are modified in a way that makes them, in technical language, non-local and non-linear. (Bohm 1987a, p. 42)

Bohm realized that the ontological interpretation of quantum field theory has some features that might help to solve one of the basic problems of the implicate order framework, namely the problem of explaining how the explicate order arises from the implicate order. This is how he reasons:

> ...classically, solutions of the field equations represent waves that spread out and diffuse independently. Thus... there is no way to explain the origination of the waves that converge to a region where a particle-like manifestation is actually detected, nor is there any factor that could explain the stability and sustained existence of such a particle-like manifestation. However, this lack is just what is supplied by the super-quantum potential. Indeed... the non-local features of this latter will introduce the required tendency of waves to converge at appropriate places, while the non-linearity will provide for the stability of recurrence of the whole process. And thus we come to a theory in which not only the activity of the particle-like manifestations, but even their actualization, e.g. their creation, sustenance, and annihilation, is organized by the super-quantum potential. (Bohm 1987a, pp. 42–43)

It is clear that the above fits well with the implicate order framework. We can think of the electromagnetic field as the "first implicate order". Of course we have already suggested something like this when in Chaps. 1 and 2 we used the way the electromagnetic field at a given small region enfolds information about the whole as an example of the implicate order. When Bohm applies the ontological interpretation to this field, the formalism suggests that there also exists an objectively real superfield that is a function of all the electromagnetic field coordinates. This superfield in the field theory is analogous to the new quantum field ψ in the particle theory. Just as the quantum field ψ in the particle theory gives rise to the quantum potential acting on the particle, the superfield in the field theory gives rise to a super-quantum potential acting on the electromagnetic field. Bohm calls the superfield "the second implicate order". When the superfield (i.e., the second implicate order) acts, via the super-quantum potential, upon the electromagnetic field (i.e., the first explicate order), the result is a photon – a momentary particle-like manifestation in a small region of space, i.e. an explicate order.

Bohm describes the idea further:

> The general picture that emerges out of this is of a wave that spreads out and converges again and again to show a kind of average particle-like behavior, while the interference and diffraction properties are, of course, still maintained. All this flows out of the super-quantum potential, which depends in principle on the state of the whole universe. But if the "wave function of the universe" falls into a set of independent factors, at least approximately, a corresponding set of relatively autonomous and independent sub-units of field function will arise. And, in fact . . . the wave function will . . . tend to factorize at the large-scale level . . . (1987a, p. 43)

This opens up a way of understanding how the everyday world arises from the underlying quantum field theoretical world:

> So now we see quite generally that the whole universe not only determines and organizes its sub-wholes, but also that it gives form to what has until now been called the elementary particles out of which everything is supposed to be constituted. What we have here is a kind of universal process of constant creation and annihilation, determined through the super-quantum potential so as to give rise to a world of form and structure, in which all manifest features are only relatively constant, recurrent and stable aspects of this whole. (Bohm 1987a, p. 43)

These manifest features correspond, of course, to what Bohm has called the explicate order. Given that Bohm was not able to develop an ontological interpretation of a quantum field theory of "fermionic" particles like electrons, the above description sounds, in my view, too optimistic, and can be misleading. The notion that all elementary particles arise from the action of the super-quantum potential should more accurately be described as Bohm's vision, the aim of the future theory, rather than as something that has already been accomplished. As we have mentioned above, there have been some interesting attempts since 1987 to apply the ontological interpretation to fermionic particles like electrons, thus making the realization of the aim at least conceivable (see, for example, Nicolic (2005)).

Bohm next describes the relationship between the super-quantum potential scheme to the implicate order scheme:

> To see how this is connected with the implicate order, we have only to note that the original holographic model was one in which the whole was constantly enfolded into and unfolded from each region of an electromagnetic field, through dynamical movement and development of the field according to the laws of classical field theory. But now, this whole field is no longer a self-contained totality; it depends crucially on the super-quantum potential. ... this in turn depends on the "wave function of the universe" in a way that is a generalization of how the quantum potential for particles depends on the wave function for a system of particles. But all such wave functions are forms of the implicate order (whether they refer to particles or to fields). Thus, the super-quantum potential expresses the activity of a new kind of implicate order. This implicate order is immensely more subtle than that of the original field, as well as more inclusive, in the sense that not only is the actual activity of the whole field enfolded in it, but also all is potentialities, along with the principles determining which of these shall become actual. (Bohm 1987a, p. 43)

As we have already mentioned, Bohm called the original field the first implicate order while he called the super-quantum potential the second implicate order (or the super-implicate order). Of course, Bohm was always keen to emphasize that reality is likely to be more complex than what our current theories imply. One way to take into account this possibility in the implicate order framework is to assume that

> there could be a third, fourth, fifth implicate order, going on to infinity, and these would correspond to extensions of the laws of physics going beyond those of the current quantum theory, in a fundamental way. But for the present I want to consider only the second implicate order, and to emphasize that this stands in relationship to the first as a source of formative, organizing and creative activity. (Bohm 1987a, pp. 43–4)

As we shall see later in this chapter, the idea of there being a series of implicate orders becomes important when Bohm tries to discuss the relationship between mind and matter in his scheme. We already saw when discussing the notion of soma-significance that Bohm appeals to the notion of levels when discussing this relationship. The ontological interpretation, and especially the notion of the super-quantum potential, provides a prototype model of how a higher level organizes a lower one.

The overall, extended notion of the implicate order described above incorporates both of the ideas that Bohm had around 1950 while in Princeton. There was the idea of the electron as a movement of outgoing and incoming waves, and the idea of the ontological interpretation of the quantum theory, including the idea of the electron as a combination of a particle and the field, and the quantum field theoretical idea of the electromagnetic field acted upon by the super-quantum potential. At first sight, these ideas, especially when it comes to the nature of an electron, may seem to be in complete contradiction. One of them accounts for the appearance of the electron as a particle in an experiment by assuming that the particle-like manifestation is a result of a wave closing into the point where the particle appears. The other one simply assumes that the electron is a particle, always accompanied and guided by a new type of field. Bohm's extended notion of the implicate order retains something about both schemes. It does not consider as fundamental the notion that an electron is a particle, although it does admit that such a notion of particle can be useful in some domains. Even in this particle model, the implicate order prevails in the wave which guides the particle. It is the ontological interpretation of fields that becomes central in the extended notion of implicate order. Indeed this interpretation provides what is missing in the original idea of the implicate order, namely an explanation of why there are incoming waves, or why the waves converge to produce a particle-like manifestation. This is explained by appealing to the super-quantum potential. So, one might say that Bohm discovered the implicate order in two different ways around 1950. The first discovery was a general, qualitative one, based on the idea that an incoming wave, in addition to the outgoing wave already described by the Schrödinger equation, would account for particle-like manifestations. The second discovery was a more specific one, accomplished by applying the ontological interpretation to the electromagnetic field. As we have seen, when this is done, then there is a super-quantum potential which brings about the tendency of waves to converge.

Thus although the idea of outgoing and incoming waves and the ontological interpretation seemed initially very different, Bohm thought that they proved to be two aspects of one more comprehensive notion:

> This can be described as an overall implicate order, which may be extended to an infinite number of levels and which objectively and self-actively differentiates and organizes itself into independent sub-wholes, while determining how these are interrelated to make up the whole. (Bohm 1987a, p. 44)

This is a succinct characterization of the notion of the implicate order, as extended through the ontological interpretation of quantum theory. The possibility of extension to infinite number of levels reflects Bohm's assumption that nature is infinite, both quantitatively and qualitatively (see Bohm (1957, Chap. 5)). There is also the idea that the universe is a self-organizing system (cf. Jantch (1980) and Kauffman (1993, 2000)).

Bohm further thinks that what he has called the explicate order emerges naturally from the underlying overall implicate order:

> ...the principles of organization of such an implicate order can even define a unique explicate order, as a particular and distinguished sub-order, in which all the elements are relatively independent and externally related. To put it differently, the explicate order itself may be obtainable from the implicate order as a special and determinate sub-order that is contained within it. (Bohm 1987a, p. 44)

Bohm's basic suggestion is that the implicate order is fundamental and the explicate order is derivative. Here is how he concludes his presentation of his general viewpoint:

> All that has been discussed here opens up the possibility of considering the cosmos as an unbroken whole through an overall implicate order. Of course, this possibility has been studied thus far in only a preliminary way, and a great deal more work is required to clarify and extend the notions... (Bohm 1987a, p. 44)

One of the things that was missing when Bohm wrote his paper was a coherent application of the ontological interpretation to the quantum field theory of fermionic particles, such as electrons. As we have mentioned, subsequent studies have moved closer to realizing this aim (see, for example, Nicolic (2005)).

We have now completed our overview of the way Bohm thought one could use the ontological interpretation to extend the implicate order framework. It is now time to consider how he tried to discuss the nature and relationship of mind and matter in this extended framework.

4.4 The Relationship between Mind and Matter in the Light of the Ontological Interpretation and the Implicate Order

We have seen in the previous section why Bohm thought the ontological interpretation can be used to extend the implicate order framework, thus making it a more specific theory about the nature of physical existence. Particularly important for this was the ontological interpretation of quantum field theory, with its notion of the super-quantum potential which provides a way of understanding how the explicate order arises from the implicate order. There remain two problems with the implicate order framework which relate to its application to mind. Firstly, it is not clear how the explicate order of conscious experience arises from the implicate order of the mind. Bohm notes that

> it is natural to suppose that the explicate space and time that we consciously experience is projected from its enfoldment in these deeper implicate orders. (Bohm 1986, p. 198)

However, he did not provide a detailed theory of exactly how it is projected. One option would be to consider Pribram's holographic theory of neural memory, and to specify how such projection could be explained in terms of this theory (see also the very interesting proposals by Marcer and Mitchell (2001)). Also, one could consider the various attempts by those who use the metaphor of virtual reality for conscious experience (e.g. Revonsuo (1995) and Lehar (2002)) to show how the virtual reality is supposed to be constructed by the brain. However, we shall not consider this very interesting issue here. Instead, we shall move on to consider the second issue concerning the mind which the implicate order framework originally left unspecific and which Bohm thought that the ontological interpretation could help to resolve: the question of how mind and matter are related to each other. We shall initially return to consider the particle theory, but shall also appeal to the field theory later on.

We shall begin by considering quantum wholeness, and the way it resembles certain biological phenomena. The idea is that when one tries to understand the relationhip between quantum theory and the mind, it is natural to consider first the relationship between quantum theory and biological phenomena. Given that mind and conscious experience, as these are known to us, are present in biological organisms, the bridge between quantum physics and biology can be a part of the bridge between quantum theory and the mind.

Bohm notes that quantum wholeness, as brought out in the ontological interpretation of quantum particle mechanics, is reminiscent of the wholeness usually associated with living organisms:

> With the quantum potential... the whole has an independent and prior significance such that, indeed, the whole may be said to organize the activities of the parts. For example, in the superconducting state it may be seen that electrons are not scattered because, through the action of the quantum potential, the whole system is undergoing a co-ordinated movement more like ballet dance than like a crowd of unorganized people. Clearly, such quantum wholeness of activity is closer to the *organized unity of functioning* of the parts of a living being than it is to the kind of unity that is obtained by putting together the parts of a machine. (Bohm 1987a, p. 38; emphasis added)

The mechanistic programme is still very much prevalent in biology. It is widely assumed that living organisms are complex machines and that the relevant functions are from the physical point of view describable in terms of classical physics, so that it is assumed that no quantum physics (or a deeper physical theory) is needed to account for life. As we saw when considering a quotation by Dennett in Chap. 1, there seems to be a faith among many researchers that typical biological phenomena such as metabolism, growth, self-repair, and reproduction can be explained in terms of the organization of the parts, or the way in which the parts work one upon another. Dennett calls this type of viewpoint mechanistic naturalism.

It is often said that "life used to be a problem, but it is now understood". It is implicit in Bohm's approach, however, that all the relevant functions and properties of biological organisms cannot be understood, from the physical point of view, in terms of classical physics. Quantum theory implies strongly that inanimate matter cannot be understood in terms of the mechanistic paradigm. This, of course, does not *prove* that living matter cannot be so understood. However, given that biological organisms clearly have "holistic" properties (consciousness being the most striking example), it seems more likely that any explanation of consciousness in biological terms will involve "holistic organizing factors" at least as complex as those found in superconductivity, and more likely a much more complex hierarchy of such factors. Quantum theory may thus in the future help us to understand the holistic features of both the living state and the conscious state of matter. The Dennettian strategy of explaining biological and mental phenomena in terms of parts interacting with each other may need to be complemented with idea that the state of the whole (e.g. common pools of information, holistic organizing factors, etc.) plays an irreducible role in these phenomena.

If quantum wholeness is similar to what we see in living organisms, active information goes even further and begins to resemble the domain of mind:

> ...the whole notion of active information suggests a rudimentary mind-like behavior of matter, for an essential quality of mind is just the activity of form, rather than of substance. Thus, for example, when we read a printed page, we do not assimilate the substance of the paper, but only the forms of the letters, and it is these forms which give rise to an information content in the reader which is manifested actively in his or her activities. A similar mind-like quality of matter reveals itself strongly at the quantum level, in the sense that the form of the wave function manifests itself in the movements of the particles. This quality does not, however, appear to a significant extent at the level at which classical physics is a valid approximation. (Bohm 1990, p. 281)

Bohm thus gives his idea about what is distinctive about mind (cf. Rorty (1979, p. 35); Feigl (1967, p. 29)). "Activity of substance" presumably refers to the Cartesian notion of extended matter in motion, pushing and pulling other material things, whereas "activity of form" is closer to the Aristotelian notion of formative and final causation, just as Bohm's notion of assimilating the forms of the letters bears a resemblance to Aristotelian hylomorphic accounts of abstraction (cf. Rorty (1979, p. 41)).

Of course, activity of form, as such, is not that "holistic" a notion. We can imagine perfectly mechanical systems (approximately describable by classical physics) which respond to form. The non-mechanistic part in Bohm's view is hinted by the statement "these forms... give rise to an information content in the reader". "Information content" is supposedly something on the "mental" side of reality, and not reducible to a purely mechanical physical process. Information content (as opposed to an "informational vehicle", or physical processes whose form encodes the information) could also be an essential part of conscious experience. Information content is presumably a result of interpreting "forms" in a given context, and the idea is that this information content "manifests" in the activity of the reader. "Meaning matters", so to speak. But how does this happen in more detail?

Bohm indeed proposes that a major part of the significance (meaning) of thought is just the activity to which a given structure of information may give rise, and he illustrates this with his favourite example, which we briefly considered above:

> We may easily verify this in our subjective experience. For example, suppose that on a dark night, we encounter some shadows. If we have information that there may be assailants in the neighbourhood, this may give rise immediately to a sense of danger, with a whole range of possible activities (fight, flight, etc.). This is not merely a mental process, but includes an involuntary and essentially unconscious process of hormones, heartbeat, and neurochemicals of various kinds, as well as physical tensions and movements. However, if we look again and see that it is only a shadow that confronts us, this thought has a calming effect, and all the activity described above ceases. Such a response to information is extremely common (e.g., information that X is a friend or an enemy, good or bad, etc.). More generally, with mind, information is thus seen to be active in all these ways, physically, chemically, electrically, etc. (Bohm 1990, p. 281)

This is a good example, for it captures the dynamic and sometimes intense nature of the relationship between mind and body, which tends to be forgotten when modern philosophers of mind, perhaps situated in big armchairs in safe offices, consider these questions. Bohm next notes that the kind of "activity of form" discussed in the above example is similar to that in connection with automatic pilots, radios, computers, DNA, and quantum processes in elementary particles such as electrons. This leads him to conclude that

> at least in the context of the processes of thought, there is a kind of active information that is simultaneously physical and mental in nature. Active information can thus serve as a kind of link or "bridge" between these two sides of reality as a whole. These two sides are inseparable, in the sense that information contained in thought, which we feel to be on the "mental" side, is at the same time a related neurophysiological, chemical, and physical activity (which is clearly what is meant by the "material" side of this thought). (Bohm 1990, p. 282)

Bohm here begins to approach the traditional problem about how the mental and the physical are related. A modern neuroscientist might say that our processes of thought are purely physical in nature.

Descartes might have said that our processes of thought are purely non-physical, mental in nature, and exist over and above the physical processes in the brain. Bohm proposes that there is something in the process of thought that is simultaneously physical and mental in nature, namely active information. Active information then connects the physical and the mental. If you like, it is the missing link between them. Information typically has a content or meaning that can be felt to be on the mental side. But that information is at the same time a related physiological activity.

Bohm points out that he has thus far considered only a small part of the significance of thought. He therefore tries to provide a more comprehensive account:

> ...our thoughts may contain a whole range of information content of different kinds. This may in turn be surveyed by a higher level of mental activity, as if it were a material object at which one were "looking". Out of this may emerge a yet more subtle level of information, whose meaning is an activity that is able to organize the original set of items of information into a greater whole. But even more subtle information of this kind can, in turn, be surveyed by a yet more subtle level of mental activity, and at least in principle this can go on indefinitely. Each of these levels may then be seen from the mental or from the material side. From the mental side, it is a potentially active information content. But from the material side, it is an actual activity that operates to organize the less subtle levels, and the latter serve as the "material" on which such operation takes place. Thus, at each level, information is the link or bridge between the two sides. (Bohm 1990, p. 282)

There are a number of ideas in the above passage. Firstly, there is the idea that information content of thoughts at a given level can be surveyed by a higher level of mental activity. When this is done, there is a change of roles, as it were. Typically, the information content of thoughts is playing the role of being something mental, in relation to the physical side of those thoughts. Changes in the information content might then result in changes in the physical side (for example, interpreting some information as meaning "assailant" may give rise to a whole range of physical activities, as we have discussed). However, when such information content is surveyed by a higher level of mental activity, it, as it were, changes its role from being something mental to something physical. This suggests that what is considered to be "mental" and "physical" is relative to the context. An information content is typically "mental" in relation to the physical side of that information, but "physical" in relation to a higher level of mental activity that may be surveying it. This could be called the relativity of the mental–physical distinction. What is mental in one context, or when viewed from one side, may be physical in another context, or when viewed from another side.

The next point is that out of the surveying may "emerge a yet more subtle level of information, whose meaning is an activity that is able to organize the original set of items of information into a greater whole." This suggests a kind of self-organization for the mind. On the one hand, there is information content contained in thoughts; on the other, there is mental activity which surveys the information content. Out of this may then emerge a new level that is able to organize the old one. But even this new level can, in turn, be surveyed by a yet more subtle level of mental activity, creating yet another level, and so on indefinitely.

The idea is that each such level then has two sides, a mental side and a physical side. When viewed from the mental side, such level is potentially active information content. At a given moment, only some of this information content is actually active, and this actual activity is seen when the level is viewed from the material side. Part of the information operates to organize the less subtle levels.

Consider the example of the shadow in the dark night. Let us think about a level of active information in the thoughts of the person. Seen from the mental side, these thoughts contain all sorts of potentially active information (such as "assailants are dangerous"). This potentially active information becomes actually active when the shadow is interpreted as the assailant. We then get an actual activity that operates to organize the less subtle levels of the nervous system, which latter serve as the "material" on which such operation takes place: "This is not merely a mental process, but includes an involuntary and essentially unconscious process of hormones, heart-beat, and neurochemicals of various kinds, as well as physical tensions and movements." So, from the mental side, a level of active information is a potentially active information content, whereas from the material side it is an actual activity that operates to organize the less subtle levels. This perhaps makes clearer what Bohm means by the mental and the physical side of a level of active information. They are inseparable but not identical. In particular, it seems that we can imagine situations where we have only a potentially active information content without any actual activity of that information.

Bohm next proposes that this relationship between the levels holds at "indefinitely great levels of subtlety". The idea is that such a possibility of going beyond any specifiable level of subtlety is the essential feature on which the possibility of intelligence is based. Intelligence for him is the ability to respond coherently to a new situation (Bohm & Peat 1987, pp. 220–227). We act always on the basis of a certain level of understanding, which is related to our tacit and explicit knowledge (i.e. information content). When we encounter a new situation, we may not be able to respond adequately from our current information content. The activity of intelligence thus typically means that we "survey" our existing mental activities and see if we can go beyond them. I believe that Bohm has this sort of an idea in mind, although he does not go into it in his 1990 paper. Such an emergence of a new level of mental activity and a consequent new level of information may well involve a "non-computable" step in the movement of mental processes, and here a connection to Roger Penrose's ideas about non-computability and intelligence could be explored (see Penrose (1989, 1994)).

At this point Bohm explores the root meaning of the word subtle, and finds there a metaphor for the brain/mind:

> It is interesting in this context to consider the meaning of subtle which is, according to the dictionary "rarefied, highly refined, delicate, elusive, indefinable". But it is even more interesting to consider its Latin root, sub-texere, which means "finely woven". This suggests a metaphor for thought as a series of more and more closely woven nets. Each can "catch" a certain content of a corresponding "fineness". The finer nets can not only show up the details of form and structure of what is "caught" in the coarser nets; they can also hold within them a further content that is implied in the latter. We have thus been led to an extension of the notion of implicate order, in which we have a series of inter-related levels in which the more subtle – i.e. "the more finely woven" levels including thought, feeling and physical reactions – both unfold and enfold those that are less subtle (i.e. "more coarsely woven"). In this series, the mental side corresponds, of course, to what is more subtle and the physical side to what is less subtle. And each mental side in turn becomes a physical side as we move in the direction of greater subtlety. (Bohm 1990, pp. 282–283)

The metaphor suggests that the brain/mind has a certain kind of hierarchical structure. It consists of a hierarchy of levels. The levels differ with respect to their subtlety. The more subtle levels unfold and enfold those that are less subtle, suggesting that the implicate order is the order that prevails in the brain/mind. The metaphor also emphasizes the relativity of the distinction between the mental and the physical. A given level can play the role of a mental level in relation to a less subtle level and that of a material level in relation to a more subtle level. There is no absolute and strict distinction between the mental and the physical, but they are different in the sense that typically the physical refers to the less subtle and the mental to the more subtle levels.

Bohm then returns to consider the quantum theory and asks what relationship it might have to the "interweaving of the physical and the mental" that we have described above:

> First, let us recall that because the quantum potential may be regarded as information whose activity is to guide the "dance" of the electrons, there is a basic similarity between the quantum behavior of a system of electrons and the behavior of mind. But if we wish to relate mental processes to the quantum theory, this similarity will have to be extended. The simplest way of doing this is to improve the analogy between mental processes and quantum processes by considering that the latter could also be capable of extension to indefinitely great levels of subtlety. (Bohm 1990, p. 283)

This means that if one wants to make concrete use of the quantum field and the quantum potential in trying to explain just how mind and matter are related, it is not enough to say that the relation of, say, the mind to the brain is analogous to the relation of the quantum field to the electron. The analogy would not yet tell us how the information content felt on the mental side is connected with the physical aspect of the information, nor how the level of information content in our thought processes would connect to the underlying levels of information, including the level of active information contained in the wave function that guides, say, the electrons and ions in the brain. Something more has to be said in order to connect mind with matter properly. Bohm has previously argued that the mind can be understood as a series of levels of information where the more subtle ones unfold and enfold the less subtle ones. His first step when connecting mind and matter is thus to suggest that matter could likewise be assumed to involve a hierarchy of levels of information, rather than just one level of active information contained in the quantum field:

> ... one could begin by supposing, for example, that as the quantum potential constitutes active information that can give form

> to the movements of the particles, so there is a superquantum potential that can give form to the unfoldment and development of this first order quantum potential. This latter would no longer satisfy the laws of the current quantum theory, which latter would then be an approximation, working only when the action of the superquantum potential can be neglected.
>
> Of course, there is no reason to stop here. One could go on to suppose a series of orders of superquantum potentials, with each order constituting information that gives form to the activity of the next lower order (which is less subtle). In this way, we could arrive at a process that would be very similar to that to which we have been led in the consideration of the relationship of various levels of subtlety in mind. (Bohm 1990, p. 283)

As we saw previously in this chapter, Bohm had actually already discovered the super-quantum potential in his 1952 papers when he applied the ontological interpretation to the electromagnetic field. We also saw in the previous section that Bohm felt it a reasonable possibility that there exists a whole series of levels, over and above the level of the super-quantum potential. Notice also that Bohm is in the above passage suggesting that we extend the *particle theory* in terms of a super-quantum potential. This is a very preliminary and schematic idea, for remember that Bohm was not able to develop a quantum field theory for particles like electrons, where the super-quantum potential would play a role in organizing the existence and behavior of the particle aspect of the electron. However, recent developments might make such extension of the particle theory a coherent possibility (see, for example, Bell (1987) and Nicolic (2005)).

Having developed the idea of two analogous hierarchies or processes, one for mind and one for matter, Bohm then goes on to consider their relationship:

> One may then ask: What is the relationship of these two processes? The answer that I want to propose here is that there are not two processes. Rather, I would suggest that both are essentially the same. This means that that which we experience as mind, in its movement through various levels of subtlety, will, in a natural way ultimately move the body by reaching the level of the quantum potential and of the "dance" of the particles. There is no unbridgeable gap or barrier between any of these levels. Rather, at each stage some kind of information is the bridge. This implies, that the quantum potential acting on atomic particles, for example, represents only one stage in the

> process. The content of our own consciousness is then part of this overall activity. (Bohm 1990, p. 283)

This is where Bohm's mind–matter theory starts making proposals that should arouse the interest of anyone who has thought about the mind–body problem. Consider, for example, the problem of mental causation. If minds are non-physical (as they at least in some sense seem to be), how could they possibly influence the body (which they obviously seem to do) without violating the laws of physics (which presumably form a causally closed system)?

Bohm's scheme suggests a new way of thinking about the issue. First of all, it suggests that minds are not completely non-physical, nor is matter completely non-mental. Instead, the mental and the physical are, generally speaking, inseparable aspects of an underlying active information. However, there is a sense in which we can say that the mental influences the physical or that the mind influences the body. The brain/mind can be thought of in terms of a hierarchy of levels of information. The more subtle levels can be characterized as "mental" and the less subtle ones as "physical". The mental levels are then able to influence the physical levels and vice versa. This is possible because, at each level, including at the quantum level, there is active information that connects the mental and the physical sides to each other. Notice, in particular, that the quantum potential acting on atomic particles provides the link between the mental and the physical that is typically missing in mind–body theories.[3]

Bohm is aware of the radical implications of his proposal:

> It is thus implied that in some sense a rudimentary mind-like quality is present even at the level of particle physics, and that as we go to subtler levels, this mind-like quality becomes stronger and more developed. (Bohm 1990, p. 283)

This reveals that Bohm accepts some kind of "panpsychism" or "panprotopsychism". Reality is deep down "psycho-physical". If we abstract a part from this reality, there will be some trace of this basic psycho-physical quality in it. If we have abstracted the "elementary particles" of physics, they are not completely physical, but demonstrate a rudimentary mind-like quality through their quantum field and the active information it contains. If we go to the other end of the spectrum and consider some very subtle states of consciousness, these will still

[3] For criticisms of the idea that quantum theoretical active information is relevant to the mind–matter problem, see Kieseppä (1997a and 1997b) and Chrisley (1997). See also Hiley and Pylkkänen (1997).

have a material aspect in an underlying neural state. There are thus "different kinds and levels of mind". Bohm notes that each kind and level of mind may have a relative autonomy and stability. What, then, is the relationship between these different kinds of minds?:

> One may... describe the essential mode of relationship of all these as participation, recalling that this word has two basic meanings, to partake of, and to take part in. Through enfoldment, each relatively autonomous kind and level of mind to one degree or another partakes of the whole. Through this it partakes of all the others in its "gathering" of information. And through the activity of this information, it similarly takes part in the whole and in every part. It is in this sort of activity that the content of the more subtle and implicate levels is unfolded (e.g., as the movement of the particle unfolds the meaning of the information that is implicit in the quantum field and as the movement of the body unfolds what is implicit in subtler levels of thought, feeling, etc.). (Bohm 1990, pp. 283–284)

When discussing the various mind–body theories in Chap. 3, we mentioned a number of different relationships that have been used to characterize how mind and matter hang together: reduction, identity, emergence, causation, and supervenience. It seems that Bohm introduces yet another relationship, namely "participation". Participation involves the key notions of the implicate order framework, namely enfoldment and unfoldment. But how do the above-mentioned, usual candidates for the relationship that is supposed to hold between mind and matter figure in Bohm's scheme, and what is their relation to the above notion of participation?

Let us consider some of these relationships, starting with reduction. Bohm does not try to reduce mind to matter, nor does he try to reduce a given level completely to another level of a set of levels (see, for example, Bohm (1986, p. 198)). Most importantly, there is no ultimate level in Bohm's theory (Bohm 1986, p. 181). To understand the significance of this, consider, for example, how Bohm thinks of the relationship of a solid object to its atomic structure:

> ...a solid object may at first be disturbed only mechanically by outside forces, but if its temperature is raised it eventually melts or evaporates. This is understood by going to a new context, in which the existence of a solid object is seen to depend on a deeper and more universal atomic structure. ... however, as long as we do not assume that there is an ultimate and funda-

> mental context, the reality of the solid is not denied. The atomic structure and the properties of the solids are both real in their contexts, which are, however, limited abstractions from a total reality that cannot be grasped in any specifiable context. (Bohm 1986, p. 187)

In this passage, Bohm uses the word context in roughly the same way as the word level is used in the mind–body debate (i.e. we can talk about the level of atoms, molecules, cells, organs, organisms, etc). It is fairly common in contemporary philosophy, especially among those who hold the doctrine of physicalism, to assume that there exists a fundamental level of physics in which strict laws prevail (see, for example, Fodor (1990)). This, then, in a sense "steals" the reality of higher levels. The higher levels are thought to be just convenient abstractions, ways of talking about what in the end is nothing but the fundamental level of physical fields and particles.

It is extremely interesting that Bohm, who was a physicist studying that alleged fundamental level of physics, actually thought that it is a mistake to assume its existence as absolutely fundamental. Instead, he felt that both the more fundamental level of physics and the higher levels (biological, neurophysiological, psychological, etc.) are "limited abstractions from a total reality that cannot be grasped in any specifiable context". He admitted that the existence of a higher-level object depends on structure at a lower level, and in this sense the lower level is more fundamental. But he assumed that even the existence of any such more fundamental level depends in turn on some other, even more fundamental level or context, and it is not reasonable to assume the existence of some level which is not context dependent in this way. Thus, it is not possible to reduce any level completely to another level or set of such levels, because these levels are themselves dependent on some more fundamental context.

Because Bohm accepts that the existence of the higher-level properties depends on the more fundamental context, he accepts some kind of supervenience of the mental upon the physical. However, unlike most philosophers of mind who appeal to supervenience, he emphasizes that the physical depends upon the the total reality which is its ground. Thus, strictly speaking, in the Bohmian universe everything supervenes on the holomovement.

Bohm discussed in his 1957 book *Causality and Chance in Modern Physics* the interesting notion of reciprocal, two-way relationships. We are very used to thinking that the macrolevel depends upon the microlevel, but not vice versa. Bohm gives some interesting examples of the way in which the macrolevel seems to have an effect upon the underlying microlevel. For example, it is common to think that temperature is identical with the mean kinetic energy of the molecules of, say, a gas. This suggests that temperature depends completely on the behavior of the molecules. Bohm asks us to consider what happens when that gas is put into different conditions. Within an oven it transforms into ions and electrons. One might now say that the temperature is identical with the mean kinetic energy of the ions and the electrons. But consider next that the gas is inside the Sun. It is still meaningful to talk about the temperature, but at the microlevel there are no longer stable particles but rather mesonic transformations. With these examples Bohm wants to draw attention to the possibility of reciprocal relationships, to situations where the macrolevel properties might have an influence upon the underlying microcontext, instead of the microcontext always determining the macroproperties.

Let us return to the issue of reductionism. If you like, Bohm is a reductionist in the sense that he assumes that in the end everything can be reduced to the "total reality".[4] However, he is not denying the reality of the things and the properties that can be thus reduced. A thing can be "real in its context", while each definable context is in the end a limited abstraction from the total reality. One important point is that Bohm assumes that this applies to any particle or field or force that physics can come up with. They are not real in some ultimate and fundamental sense (thus constituting the fundamental level of reality). Instead they, too, are relatively real, real in their contexts. There is thus a sense in which the fundamental particles and fields of physics and, say, psychological properties have the same ontological status. Both are real in their contexts, but these contexts are abstractions from the total reality.

What does it mean to say that something is real in its context? The traditional notion of a substance holds that a substance is ontologically independent in the sense that it can exist even if nothing else exists. Bohm is saying that all things that we can know (both mental and material things) are real, but only in their own context. Thus, they are not, strictly speaking, substances. Note especially that matter is not a substance in Bohm's ontology. It is only the "total reality" or the holomovement that can exist independently, and there is thus only one substance in the Bohmian universe. Mind cannot be fully reduced to matter; rather both mind and matter can be, in a sense, reduced to the

---

[4] I thank Otto Lappi for pointing this out to me in correspondence, 26 October 2004.

total reality. Bohm says that the "total reality... cannot be grasped in any specifiable context". This means that it cannot be known completely. It also means that it is not reasonable to think about the total reality as a "level" or "context". It is, rather, a kind of "transcendental" ground which makes it possible for there to exist "levels" and "contexts", as relatively autonomous sub-totalities of the total reality, sub-totalities which we can mentally or conceptually abstract.

How about "emergence"? Bohm is not an emergent materialist, in the sense that he would assume that matter is the fundamental substance of the universe, out of which mind has emerged at some point in evolution. However, there seem to be other ways in which he appeals to something like emergence. He assumes that reality is deep down "movement", and because the implicate order prevails in this movement, there is the possibility of an explicate order arising, as a relatively stable, autonomous context, whenever the unfolding–enfolding processes are sufficiently recurrent. There is a thus a sense in which the explicate order emerges out of the holomovement. Bohm here also appeals to the notion of potentiality. There exist potentialities in the holomovement, and things emerge as these potentialities are actualized.

In philosophical discussions the notion of emergence is often used to refer to situations in which there is no way of explaining how the emergent property arises. For example, the more extreme emergent materialists hold that there is no way of understanding how the mind emerged from matter, thus making this viewpoint, for all practical purposes, a kind of property dualism. Sometimes Bohm seems to appeal to an emergence of this kind. For example, we saw in Chap. 2 that when discussing biological evolution, he characterized creativity as the "inception of new content" and suggested that in biological evolution there appear forms that are not completely explainable in terms of (or derivable from) the previous forms in previous moments. In this sense, it seems that the new forms just emerge in a strong, unexplainable sense. However, Bohm is not saying that there is no way of understanding how they emerge. It might be possible to understand this, but to do this it is not sufficient merely to study the content of the previous moments in a particular line of history of biological evolution. We need to consider a deeper context, the multi-dimensional ground from which the moments are projected in the first place. Like everything else, these forms arise from the holomovement, and thus their origin might be understood at least up to some point by studying the implicate orders that underlie them.

Finally, we saw in this chapter that Bohm says that a new level of active information in thought can arise, when a higher level of mental activity surveys the information contents of thoughts at a given level. This can also be seen as a kind of emergence.

What about causality? Searle, in particular, has advocated the view that "brains cause minds" – that the physiological processes at the microlevel cause the mental properties at the macrolevel, a bit analogously to the behavior of molecules at the microlevel giving rise to macrolevel properties (e.g. solidity) (see, for example, Searle (1992)). He is happy to call this "vertical causation" or instantaneous causation. He also seems happy to admit that to say that some neural process causes a mental property is equivalent to saying that this neural property is instantiated by the neural process.[5] To what extent does causality play a role in the Bohmian mind–matter scheme? We saw in Chap. 3 that in his first attempt to discuss the relationship between mind and body in the implicate order framework, Bohm proposed that they have a "non-causal" relationship, analogous to the "non-causal" relationship of non-locally correlated quantum systems. The proposal was that mind and body are lower-dimensional, correlated projections from a higher-dimensional ground. Of course, we could say that it is the higher-dimensional ground that is the cause of the correlation between mind and body, and in some sense the underlying cause of the existence of both mind and body as relatively autonomous sub-totalities. When Bohm extends the implicate order framework with the help of the ontological interpretation of the quantum theory, it seems that it becomes possible to apply the notion of causality in an even more comprehensive way. As we have seen, in this extended framework Bohm thinks of the brain/mind in terms of a series of levels of active information. The levels are related via unfoldment and enfoldment, in the sense that the more subtle levels unfold and enfold the less subtle ones. Thus changes in the less subtle levels can affect the more subtle ones, and vice versa. This can be understood as a two-way causal influence between "mind" and "body", assuming that the more subtle levels correspond to "mind" and the less subtle to "body".

Bohm's notion of "participation" between the different levels of the brain/mind is thus not a reductionist notion. It suggests that there can be a strong two-way dependence between the levels, thus making it look somewhat similar to suggestions about "two-way supervenience" in philosophy of mind (see, for example, Miller (1990)). It also suggests

[5] John Searle, private communication, in Karlovy Vary, Czech Republic, August 1993.

that there can be a kind of two-way causal relationship between the levels. However, this should not be understood as a causal interaction between two separate substances.

The notion of participation has also been used to characterize the nature of measurement at the quantum level (e.g. Wheeler), and it seems to me that Bohm's notion of participation in the mind–matter context is similar to the notion of participation in the quantum context. In a typical measurement at the quantum level, we do not check a pre-existing property of the observed system without influencing it. On the contrary, the measuring apparatus and the observed system typically participate in each other. They form an undivided whole. At the end of the interaction, the measuring apparatus will be left into a certain state, indicating that the observed system had that value of that property when the interaction ended. However, the measuring apparatus took part in producing that result, and it is typically not possible to say that one has measured a pre-existent property of the observed system. This suggests that the new ontological features at the quantum level, such as quantum wholeness, have important epistemic implications, or implications for our possibility of obtaining knowledge about quantum reality.

As I mentioned, it seems to me that Bohm thinks of the relationship between the various levels of the brain/mind as analogous to the relationship between the measuring apparatus and the observed system at the quantum level. There is a radical kind of mutual participation between the different levels of the brain/mind, suggesting a strongly holistic view of a human being:

> For the human being, all of this implies a thoroughgoing wholeness, in which mental and physical sides participate very closely in each other. Likewise, intellect, emotion, and the whole state of the body are in a similar flux of fundamental participation. Thus, there is no real division between mind and matter, psyche and soma. The common term psychosomatic is in this way seen to be misleading, as it suggests the Cartesian notion of two distinct substances in some kind of interaction (if not through the action of God, then perhaps in some other way) (Bohm 1990, p. 284)

Bohm thus suggests that it is participation (rather than reduction, emergence, causation, or supervenience) that is the key relationship that prevails between mind and body, although he does not deny that the other relationships play some role as well. Bohm's view is trying to be holistic, and he tries to accomplish this wholeness without reducing mind to matter. Of course, one might say that the materialist viewpoints which are common in contemporary philosophy of mind are also aiming for a certain kind of wholeness or unified view, in their denial of Cartesian substance dualism. However, many philosophers admit that materialism does not do justice to the mental: it leaves it out as it were, and of course, this amounts to a fragmentation of the human being. Bohm is trying to have a holistic view of mind and matter without falling into the extremes of dualism and materialism. I think he has chosen a promising track, but, of course, his scheme is fairly complex and needs careful reflection and further development.

Bohm extends his holism beyond the individual human being:

> Extending this view, we see that each human being similarly participates in an inseparable way in society and in the planet as a whole. What may be suggested further is that such participation goes on to a greater collective mind, and perhaps ultimately to some yet more comprehensive mind in principle capable of going indefinitely beyond even the human species as a whole. (This may be compared to some of Jung's... notions.) (Bohm 1990, p. 284)

This passage echoes Bohm's early discussion in the implicate order framework, which we considered in Chap. 3. Bohm used to describe culture as "shared meaning", suggesting that there is an analogy between society and the notion of "common pool of information" which plays a radically holistic role in quantum many-body systems (e.g. in superconductivity). Indeed, one of Bohm's long-term interests was communication. As a result of a contact with the London-based group therapist Patrick de Mare, he became interested in a certain kind of discussion in a large group of people, which de Mare called "dialogue". The word dialogue is here not meant as a discussion between two people, but rather in its etymological sense of *dia* ("through") + *logos* ("meaning"), suggesting that dialogue involves a flow of meaning through the minds of a group of people. I mentioned in Chap. 2 that Bohm was very concerned with the social implications of our general world view, and with understanding the source of the various problems of humanity. He felt that a major source of these problems is in our habits of thought, which are overly mechanistic, divisive, and egoistic. He saw that one basic trouble was the tendency of the human mind to hold onto non-negotiable positions and a kind of self-deception which covered up the unreasonableness of such holding. Encouraged by de Mare's work, he thus engaged in a number of large group discussions or dialogues, both during shorter periods such as weekends and with groups meeting for a longer period of time on a weekly basis. He outlined certain principles whose aim was to facilitate the emergence of a flow of meaning in the group, the birth of a true microculture, or shared meaning. These kinds of activities are known as "Bohmian dialogue", and have aroused a fair amount interest among organization theorists (Senge 1990), and some universities (e.g. the MIT dialogue project; see Isaacs (1999)).

The underlying idea in the Bohmian project is that a new general world view, and more specifically a new, more adequate view of issues such as the mind–matter relationship, meaning, and language might give rise to a new kind of operation of the mind, with new, coherent ways of thinking. Although these issues are not the main subjects of this book, I find them very important and interesting. Unfortunately, science and philosophy have a tendency to become fairly mechanical affairs, driven by various more or less superficial interests, not the least various economical and technological ones. Even in these fields true communication and a genuine search for truth are fairly rare. Even if one might not always agree with Bohm's particular views about the source of the problems of society and the way to approach them, I think the general line of approach he opened up is well worth exploring, and it is indeed urgent that something like this is done. For one thing it might help science and philosophy to better achieve the aims they have set for themselves.

What might Bohm mean when he says that the human being participates "in the planet as a whole"? Here one might think of Lovelock's Gaia hypothesis, the idea that the planet as a whole can be seen as a kind of living organism. The mechanistic tradition in science and philosophy has emphasized the idea that human beings stand in an external relation to physical things, including the Earth. The implicate order scheme, when extended with the ontological interpretation of quantum theory, draws attention to the possibility of internal relationships and participation as an example of such a relationship. Of course, when one thinks of this, it is fairly obvious that human beings truly participate in the life of the planet Earth (instead of just interacting with it without significantly influencing it). So-called primitive societies have typically had very powerful ways of seeking contact with nature, not just intellectually, but in various "aconceptual" ways (cf. Pylkkö 1998). Science and philosophy have made possible another kind of participation. Scientific knowledge of the laws of nature enables the control of natural processes in a very powerful way and an advanced technology. Of course, we know all too well the dangers of this kind of participation. There is, for example, the risk of nuclear war and various kinds of environmental crisis. No doubt this prompted scientists like Bohm and his good friend the biologist Maurice Wilkins to seek for ways in which scientific knowledge could be applied without destroying the planet. This brings us back to the issues we discussed in the previous paragraph, namely the importance of understanding the factors which lead human beings to a failure of communication and coherent action.

What are we, finally, to make out of Bohm's reference to a "greater collective mind" and a "mind . . . capable of going beyond . . . the human species as a whole". The "collective mind" is something Bohm explored in his experiments with the large group, or dialogue. He used to emphasize the essentially social, collective nature of the thinking process. In a large group dialogue, this collective nature may reveal itself more powerfully than in everyday experience, which may be dominated by thought as the "inner speech" in the mind of the individual, and the interaction between such individuals. I used to take part in many of Bohm's dialogue experiments and could witness the at times very powerful nature of the "group mind". Bohm's dream was the awakening of a kind of "collective intelligence" that a group mind could exhibit. This idea sounded a bit totalitarian to me, and I challenged Bohm, asking whether such a mind would really be a desirable sort of thing. I recall him answering that he thought the individuals taking part in such a common mind would not lose their critical faculties or own intelligence but would be able to see illusions more effectively. Thus it seems he felt that such a "collective intelligence" would make possible more freedom even for the individual.

I think by "collective mind" Bohm also thought of something more general than a group mind that might emerge in a dialogue group. Presumably the idea here is that a species might, at some level and in some sense, have a common mind. Do ants have minds, and do they have a common mind, given the very organized behavior they carry out? Does humanity, in some sense at least, have a common mind, made possible, for example, by the various mass media? I will not explore these questions here in more detail, though they surely are interesting and potentially important ones to explore.

What about the mind allegedly going beyond the human species? In the framework of mechanistic science, it is fairly difficult to make sense of this type of idea. However, the implicate order framework suggests that all kinds and levels of mind are in the end projections from a higher-dimensional ground. We considered in Chap. 3 Bohm's suggestion that consciousness might have its ground in a kind of energetic emptiness, analogous to the immense sea of energy in empty space. He also suggested that this energy, the ground of consciousness could be sensed in some way as "emptiness". No doubt these suggestions echo his many intense discussions with the Indian-born teacher J. Krishnamurti, which were very important to him for a long period of time, from the early 1960s until Krishnamurti's death in 1986 (see, for example, Krishnamurti and Bohm (1999) and Peat (1996)).

Bohm moves on to consider a potential problem in his scheme:

> Finally, we may ask how we can understand this theory if the subtle levels are carried to infinity. Does the goal of comprehension constantly recede as we try to do this? I suggest that the appearance of such a recession is in essence just a feature of our language, which tends to give too much emphasis to the analytic side of our thought processes. (Bohm 1990, p. 284)

To explain what he means, he considers yet another analogy, that of the poles of a magnet (Fig. 4.5). These poles, he suggests,

> are likewise a feature of linguistic and intellectual analysis, and have no independent existence outside such analysis. . . . at every part of a magnet, there is a potential pair of north and south poles that overlap each other. But these magnetic poles are actually abstractions, introduced for convenience of thinking about what is going on, while the whole process is a deeper reality – an unbroken magnetic field that is present over all space. (Bohm 1990, pp. 284–285)

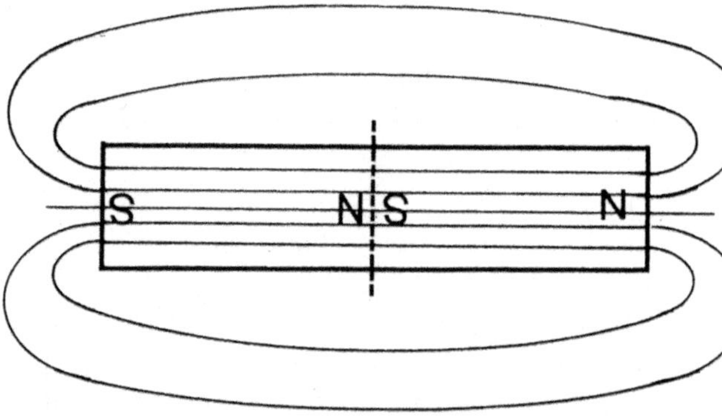

Fig. 4.5. Magnetic poles as abstractions from an overall magnetic field

Thus, if one actually breaks a magnet, one gets two magnets. But we are free to choose at which point we break it. In this sense there exists potentially both a north pole and a south pole at each point of a magnet. This is the essence of the metaphor Bohm suggests for mind and matter:

> Similarly, we may for the sake of thinking about the subject abstract any given level of subtlety out of the unbroken whole of the reality and focus our attention on it. At each such level, there will be a "mental pole" and a "physical pole". Thus... even an electron has at least a rudimentary mental pole, represented mathematically by the quantum potential. Vice versa... even subtle mental processes have a physical pole. But the deeper reality is something beyond either mind or matter, both of which are only aspects that serve as terms of analysis. These can contribute to our understanding of what is happening but are in no sense separate substances in interaction. Nor are we reducing one pole to a mere function or aspect of the other (e.g. as is done in materialism and in idealism). (Bohm 1990, p. 285)

The issue Bohm is struggling with above is, of course, one of the basic issues: what is the relationship between our representations and the reality they allegedly represent? Bohm's theory seems to have certain problems, such as the infinite regress of levels. However, he suggests that this is a feature of the description, not of reality itself. He assumes that, strictly speaking, reality is an unbroken whole. However, it is in our interests try to understand this reality. To do this, we have to abstract some parts of reality, as we cannot deal with the unbroken whole of reality. There is thus a difference between our abstractions and the reality out of which the abstractions are made. The idea is that we have to be careful not to attribute too strong a reality to our abstractions. For example, Bohm agrees that it is useful to abstract levels out of reality, thus suggesting that the notion of "level" is a useful abstraction. However, it is a feature of the abstracting process that it is in principle possible to abstract an infinite number of levels, suggesting that we will never be able to understand reality. Here Bohm seems to suggest that we must not take the idea of the inifinity of levels too literally or seriously; the appearance of the infinity of levels is more the property of the abstractions that of reality itself. Indeed, Bohm suggested in his 1957 book *Causality and Chance in Modern Physics* that although it is very useful to think of nature in terms of a hierarchy of levels, there may come a point at which the notion of levels no longer applies and something other is needed. For example, as we go to smaller and smaller spatial scales, the notion of levels consisting of smaller and smaller entities may have to be replaced by a notion of a "background", relating to the fields which underlie the elementary particles.

Similarly, Bohm admits that it is useful to abstract the distinct poles "mental" and "physical" when considering a given level that has been abstracted. If we are considering human beings, it is of course natural to make such an abstraction. The trouble, however, is that there has been a tendency in philosophy to forget that the "mental" and the "physical" are useful abstractions, and instead attribute independent reality to them. In the case of Descartes, this led, of course, to substance dualism. However, something similar may be going on even in contemporary philosophy of mind. In particular, philosophers tend to forget that the "physical" is in the end an abstraction. Instead, they attribute independent reality to the physical and are then led to worry about the status of the "mental", which is no longer given an equally strong ontological status. This then gives rise to the strong urge to "reduce" the mental to the physical. However, these attempts are often very unsatisfactory, leaving one with the feeling that contemporary philosophy is basically incapable of dealing with the relationship between the mental and the physical in a coherent way.

It is interesting that Bohm, who is a physicist, is strongly urging the scientific and philosophical community to realize that the "physical" is also an abstraction – a useful abstraction, but in the end only an abstraction. It is only "Reality" that has independent reality, while abstractions, being abstractions from Reality, have only a relative reality: they are real in their own contexts. In particular, they are not substances.

Bohm finally draws attention to the crucial importance of the ontological interpretation of quantum theory to mind–matter theory:

> The key point is... that before the advent of the quantum theory, our knowledge of matter as gained from the study of physics would have led us to deny that it could have a mental pole, which would enable it to participate with mind in the relationships that have been described here. We can now say that this knowledge of matter (as well as of mind) as changed in such a way as to support the approach that has been described here. To pursue this approach further might perhaps enable us to extend our knowledge of both poles into new domains. (Bohm 1990, p. 285)

In some ways Bohm's strategy seems similar to that of the eliminative materialists. For Bohm is appealing to the development of science to support his viewpoint. The traditional view of matter is that is has no "mental pole", making it very difficult to understand its relationship with mind. This view of matter, which derives from classical physics, has become part of the common-sense view of matter, so much so that we might call it "folk physics". Now, remember that the eliminativist materialists suggest that our common-sense theories about the mind and consciousness form a kind of theory that can be called "folk psychology". They further think that scientific developments show that folk psychology is a mistaken theory. Thus there is no need to reduce it, but rather we should just give up or eliminate those parts which are in contradiction with scientific psychology and neuroscience. Analogously, we could say in the light of Bohm's views that parts of folk physics are simply mistaken and should be given up or eliminated. In particular, we ought to question the common assumption of folk physics, namely that matter has no mental pole. In the light of Bohm's views, a more scientific notion of matter should attribute a rudimentary mental pole to matter, which opens up a new way of understanding the relationship of mind and matter.

Now, I am aware that many people will not be convinced by the above suggestion. Perhaps the idea of matter having a rudimentary mental pole goes so strongly against both our scientific and philosophical tradition, and our folk physics, that most educated people will simply dismiss it as absurd. However, some of the best contemporary philosophers, like Thomas Nagel, have pointed out that some sort of panpsychism seems to be implicit in some fairly common contemporary materialist approaches to the mind (see Nagel (1978, Chap. 13). At the same time, Bohm suggests that physics strongly supports some kind of panpsychism, and a number of other quantum physicists have made similar suggestions. So my suggestion is that although the idea of matter having proto-mental properties is admittedly bizarre (I have myself found it very difficult to digest), it may be worth thinking about seriously (see also Pylkkänen (1996a)).

Of course, it is important to realize that the way in which one ought to define panpsychism in the Bohmian scheme may be somewhat different from other ways of defining it. Consider, for example, how Noordhof defines panpsychism in *The Oxford Companion to Philosophy*:

> **panpsychism** is a doctrine about the nature of spatio-temporal reality. It asserts that each spatio-temporal thing has a mental or "inner" aspect. Few panpsychists would be happy with a characterization of their view as that all things have minds, even sticks and stones. Instead, they want to say that there

> may be varying degrees in which things have inner subjective or quasi-conscious aspects, some very unlike what we experience as consciousness. A full-blown mind would only be possessed by things approaching the complexity of human beings. On the other hand, it is difficult to characterize precisely to what extent all spatio-temporal things are supposed to have an inner "mental" aspect. Most of those who espouse this doctrine feel impelled to do so because they do not see how the mental can be caused by, or composed from, non-mental things. (Honderich 1995, p. 641)

In Bohm's ontological interpretation of quantum particle mechanics, a quantum particle, such as an electron, is a spatio-temporal thing, and it also has a "proto-mental" quality in its field of active information. It is very important to underline the potential significance of this. Panpsychism is an old philosophical doctrine with roots in Spinoza and Leibniz. Bohm is providing some support for it from modern science.

But note also that Bohm does not think that there exists a material substance and that each part of that material substance has a proto-mental property. Strictly speaking, panpsychism, in the Bohmian scheme, is just a way of characterizing our abstractions. We can abstract different levels from the total reality, and whenever we do this, Bohm suggests we will find both a mental and a physical pole. Or more accurately, we will find that the concepts "mental" and "physical" will be useful tools when we try to understand the nature of the levels we have abstracted, and through that, the total reality beyond such levels.

Thus, although I think it is interesting to consider Bohm's scheme as involving a kind of panpsychism, it is important to realize that it is a rather weak form of panpsychism. Also, I recall him not liking the term panpsychism, as the reference to "psyche" brings to mind a separate entity. I suggested the term "pansignifism", in an attempt to capture Bohm's idea that it is, strictly speaking, "significance" that we find at each level. Although he had a taste for neologisms, for some reason he did not warm to this particular one.

Note finally how, according to Noordhof, panpsychism is a "doctrine about the nature of spatio-temporal reality". Of course, the Bohmian scheme does not attribute such a supreme value to spatio-temporal reality as is usually done in contemporary philosophy. Bohm was drawing attention to the implicate order, which is also the ground of space-time. The deeper essence of consciousness might thus be beyond space-time. Thus, one could try to define Bohmian panpsychism more holistically, for example by suggesting that consciousness is enfolded in the holomovement but does not manifest in all systems in the explicate order, but only in those with sufficient complexity and subtlety (for example, that of mammalian brains). This, of course, is consistent with our everyday experience.

5

Time Consciousness

5.1 Introduction

Conscious experience has become a focus of intense study in recent years in philosophy, psychology, cognitive neuroscience, cognitive science, artificial intelligence, etc., so much so that it is fair to say that a new interdisciplinary field called consciousness studies has been born (see, for example, Güzeldere (1997)). An important aim of consciousness studies is simply to explain the various puzzling aspects of consciousness. Following van Gulick's (1995) classification, the puzzling aspects that need explanation include (a) the difference between conscious mental states and nonconscious or unconscious mental states or processes; (b) the distinction between conscious and non-conscious or unconscious creatures; (c) qualia and the qualitative nature of conscious experience; (d) subjectivity; and (e) intrinsic intentionality/semantic transparency. There is yet another feature which van Gulick picks up on. While consciousness researchers sometimes use the term "phenomenal" interchangeably with "qualitative" (connected to "raw feels"), van Gulick prefers to reserve "phenomenal" for a more comprehensive range of features:

> Current philosophical debate has focused heavily on raw feels, but they are just one aspect of our experienced inner life and thus only part of what we must deal with if we aim to describe *the phenomenal structure of experience.* In this sense the use of "phenomenal" accords better with its historical use by Kant and later by the phenomenologists. The order and connectedness that we find within experience, its conceptual organization, its temporal structure, its emotive tones and moods, and the fact that our experience is that of a (more or less) unified self set over

against an objective world are just a few of features other than raw feels that properly fall within the bounds of the phenomenal. All will need to be addressed if we take the phenomenal aspect as our explanandum. (van Gulick 1995, p. 64)

In this chapter we will be concerned with trying to explain and understand some aspects of the phenomenal structure of experience in a new way. In particular, we are concerned with the temporal structure, such as understanding our experience of movement or the sense of flow in our experience. How does this latter arise? What is its relationship to the underlying neural processes? What is the relationship between our immediate experience of movement to our attempts to model movement in terms of our conceptual thought and mathematical models? And what is the relationship between our experience of movement in conscious experience and movement in the physical domain? As we saw in Chap. 3, these questions were discussed in an original way by David Bohm, especially in the last chapter of his 1980 book *Wholeness and the Implicate Order* (see also, for example, Bohm (1965c)). His views about time consciousness, however, have thus far been by and large ignored in the consciousness studies community. One of the main arguments of this chapter is that they are very useful tools for anyone who wants to understand the temporal aspects of conscious experience. But I want to motivate the consideration of Bohm's views by first examining some prominent past and recent attempts to tackle the problems of time consciousness. These views are interesting, but as I will try to show, they suffer from certain problems. The next step will then be to show that these problems, at least to some extent, can be resolved or avoided in Bohm's scheme.

In the contemporary debate, a very important contribution to the understanding of time consciousness has been made by the philosopher Barry Dainton. In his two recent books *The Stream of Consciousness* (Dainton 2000) and *Time and Space* (Dainton 2001), Dainton has focused on understanding the temporal features of our experience, or the character of our experience from moment to moment within our "streams of consciousness". Dainton's approach is *phenomenological*, in the sense that his aim is to formulate a description of our short-term experience of time that is clear, accurate, and intelligible, as opposed to, say, trying to offer an explanation of such experience in terms of neural processes.

Antti Revonsuo (2003) has in a recent commentary not only praised Dainton's approach for its phenomenological acuteness, but also criticized it for failing to offer us genuine explanations for the unity and continuity of consciousness. The origin of the trouble seems to be Dainton's descriptive method, which means that he stays at the level of phenomenal content and is not sufficiently concerned with other levels of description, such as the neural level, when trying to understand phenomenal temporality. Revonsuo writes

> I believe that many empirically minded scientists share my view that a genuine explanation goes beyond the level of description where the *explanandum* itself is identified.

In this chapter I will first give a short summary of what Dainton is trying to explain, as well as of his description of some previous models of time consciousness and their problems. I will then present Dainton's own view, or the simple overlap theory. The next step will be to briefly consider Revonsuo's comments on Dainton's view and to point out that I agree with both Revonsuo's praise of Dainton, as well as parts of his critique. I will also offer additional criticisms of Dainton's view. I will then move on to consider another approach to time consciousness which was proposed by Bohm in *Wholeness and the Implicate Order* (Bohm 1980), and which we have discussed in Chap. 3. Bohm's model can be seen as a certain modification of the so-called two-dimensional model which Broad and Husserl, for example, held. In my view, Bohm's model seems to be able to avoid some of the problems of both the simple overlap model and the two-dimensional view, while retaining some of the explanatory value of the two-dimensional model. Note also that it is essential in Bohm's scheme to try to understand our experience of time consciousness in relation to the underlying neural processes. This suggests that Bohm's model is not vulnerable to the sorts of objections which Revonsuo raised against Dainton.

5.2 What is Dainton Trying to Explain?

Dainton's concern is the character of our experience from moment to moment within our streams of consciousness. His aim is to provide us with a clear, accurate, and intelligible description of our short-term experience of time. Central to him is the observation that our typical experience is a combination of both persistence and change. He thus takes it as obvious that consciousness is alive with change and variation. He notes that change is something that we see, hear, feel, and imagine. There can also be changes in what we smell and taste and in our emotional state. And there is also change in thought, as we can witness in our inner speech.

Dainton also wants to draw our attention to

> the subtle but distinctive sort of dynamism that is characteristic of *un*changing sensations. (Dainton 2001, p. 94)

Even when nothing seems to be changing in our sensations, there is a passive awareness of continuation and renewal (e.g. hearing a continuous and continuing flow of sound, like a note played on a cello). He calls such a feature "immanent phenomenal flow", and suggests that it is possessed by all forms of experience. There is a constant and constantly flowing presence within our consciousness.

There are thus two facts which Dainton thinks that any adequate account of the temporal features of experience must accommodate. Firstly, he maintains that our experience of change is as direct and immediate as our experience of colour and shape. Secondly, he maintains that our experiences possess the feature of phenomenal flow. Dainton sets out to give an account of phenomenal temporality, and in particular he wants to answer the following questions:

- It seems obvious that we are directly aware of change, but how is this possible?
- It seems obvious that we are directly aware of endurance, or phenomenal passage, but how is this possible?
- It seems that the specious present is of short duration, but why would this be so?[1]
- It seems that consciousness flows in a particular direction, but why is this?

Before considering how Dainton himself proposes to answer these questions, let me give a brief summary of his very useful account of previous attempts to deal with time consciousness, as well as their problems, which he provides in Chap. 7 of his *Time and Space.*

5.3 Dainton on Previous Accounts of Phenomenal Temporality

Dainton considers a very simple example to illustrate phenomenal temporality. He asks us to consider a fast sequence of notes played on a piano: C–D–E–F-etc. The idea is that we hear each note as possessing a short but discernible duration, but we also hear the succession:

[1] Le Poidevin (2004) characterizes a common way to understand the term "specious present" succinctly as follows: "The experienced present is 'specious' in that, unlike the objective present, it is an interval and not a durationless instant."

C-being-followed-by-D, and D-being-followed-by-E, etc. His strategy is then to consider how well the various previous accounts of time consciousness manage to explain this simple experience.

There are, first of all, accounts which suggest that while we hear a note, say D, we also remember hearing C, etc. Dainton rejects such "memory-based accounts" as implausible, even when modified to involve beliefs about when a note was heard (2001, pp. 95–96).

Another option is to assume that our streams of consciousness consist of short *pulses* of experience, each of some finite duration. If within a pulse P_1 we hear C–D and in the next pulse we hear E–F, we could account for the fact that we experience not just individual tones but also successions. But Dainton rejects the pulse theory because if it were correct, we should be able to discern two distinct forms of transition within our consciousness: change within pulses and change between pulses (e.g. C–D, D–F). But we do not notice such a thing, for we experience every transition between parts of the scale in the same way. Also, the idea of separate pulses fails to capture something that Dainton assumes to be an important fact: the feature that each phase of a typical stream of consciousness is *phenomenally bonded* to its immediate predecessor and successor. He assumes that there is a direct experiental link between adjacent co-streamal phases ("co-streamal" here means that the phases are part of the same stream of consciousness). He thinks that any adequate account of temporal experience must have room for such *phenomenal bonding*.

To understand the next approach (the "awareness-overlap model") Dainton considers, it is useful to make a distinction between succession of experiences (e.g. hearing C and then D and then E) and having an experience of succession (hearing C-being-followed-by-D). For Dainton it is clear that we have experience of succession rather than mere successions of experiences. But how can we explain this? One approach is to assume that there is a distinction between phenomenal contents and acts of awareness. Experience of succession could then be understood as an act of awareness that embraces a succession of momentary phenomenal contents. For example, the act of awareness A_1 could embrace tones C and D, and the subsequent act of awareness A_2 could embrace E and F, etc. The trouble is that because there is no experienced connection between D and E, this is the pulse theory in another guise. Dainton notes that the remedy (advocated by Broad (1923)) is to assume that the scope (and so the contents) of the two acts overlapped, so that D was experienced in both A_1 and A_2. One trouble with this model is that it implies a fragmented account of our streams of consciousness.

The acts themselves are assumed to be wholly distinct, even though they provide awareness of the same contents. There is partial unity at the level of content, but no unity or continuity whatsoever at the level of awareness. The more obvious trouble is that the model implies that a phenomenal content (e.g. a tone D) which falls within the region of overlap should be heard twice, when in actual fact it is heard just once. This is the "repeated contents" problem.

Dainton then moves on to consider a very different account of phenomenal temporality, which he calls the "two-dimensional model", developed by Broad in later years (and similar to one of Husserl's models). The basic idea here is that each momentary act consists of an awareness of a different momentary content together with representations of the preceding contents. Past contents are not presented again as they initially occurred, but rather under a distinctive temporal mode of presentation. Broad introduced here a notion of "presentedness". When a content is initially experienced, it possesses this quality to the highest degree, as it is re-presented with subsequent contents it possesses less and less of this quality, until in the end it drops out of awareness altogether. In a typical specious present, the subject's span of direct awareness is filled with contents of varying degrees of presentedness. The dynamic character of experience is explained by positing a two-dimensional time: there is the earlier–later ordering of acts of awareness, and the earlier–later ordering of contents presented to these acts. Dainton notes that according to the two-dimensionalist, the specious present consists of a momentary awareness with a complex content that *seems* temporally extended. The appearance of a future-directed motion is a consequence of the ways in which the complex contents of different acts are interrelated; for example, tone C is represented in the contents of successive acts of awareness, but seems increasingly past in each. The model explains how we can have an immediate experience of both change and phenomenal flow (the latter being the impression we have that a qualitative unvarying sensation is continuously refreshed and renewed).

Dainton criticizes the two-dimensional model, firstly by saying that it is phenomenally inaccurate. The model predicts that experiences would never end abruptly, but that they would always linger on for a short while as they are represented in successive awarenesses as possessing diminishing presentedness. Also, at each moment we would be aware of our current experience together with a constantly shifting complex of representations of recent experiences, and so our consciousness would be choked with the residues of recent experiences. Dainton says that this is wrong on both counts. Our consciousness is not clogged up with fading residues of prior experiences, and experiences do not always linger; they sometimes do end abruptly. It doesn't help to say that the residues and shifting constellations of representations are very brief and thus unnoticeable. According to Dainton, if we cannot discern the posited representations, there is no reason to suppose that they have any phenomenological reality, and hence no reason to suppose they contribute to our actual experience of change and persistence.

Dainton's second criticism is that the notion of presentedness is problematic. How should one make sense of the property of presentedness that creates the extra dimension? This is (presumably) an intrinsic property, the intensity of which can vary, and which is responsible for an experience's seeming to be actual, or just past, or slightly more past, and so on. But, he asks, what quality, added to, say, a pain sensation, would make that sensation seem to be in the past rather than in the present? If I am now experiencing some sensation, won't the experience seem to be occurring now irrespective of any peculiar qualities it might possess?

Dainton's third criticism is that the two-dimensional model is profoundly atomistic, and thus implies that consciousness is fragmented. Each momentary awareness constitutes a distinct episode of experiencing in its own right, and so, from a purely experiental view, each awareness is entirely isolated from its immediate neighbours. According to the two-dimensionalist, successive acts apprehend numerically distinct contents. The model thus suffers from the same problem as the pulse account: it fails utterly to accommodate the reality of phenomenal binding. Dainton concludes that since it is obvious that our experience is not confined to isolated momentary capsules, we must look elsewhere for a realistic account of time consciousness.

In summary, Dainton criticizes the model for its phenomenal inaccuracy, the problem with the notion of presentedness, and for the fragmentary view of consciousness.

5.4 Dainton's Account of Phenomenal Temporality

Dainton then goes on to develop his own favourite view, the (simple or one-level) overlap theory. He first reminds us of the repeated contents problem of the awareness-overlap model, and goes on to consider Foster's simple solution to it. Foster's idea was to assume that it is the acts of awareness themselves that partially overlap. Thus if A_1 is an act of awareness with content C–D and A_2 with D–E, we now say that these acts are not distinct, but that they overlap partially. The idea is that the part of A_1 which apprehends D is numerically identical to the part of A_2 that apprehends D. This seems to solve the problem of repeating contents.

Dainton next wants to give up the distinction between phenomenal contents and acts of awareness altogether. The idea is that phenomenal contents are *intrinsically conscious* items. They do not need to be apprehended by a separate act of awareness in order to be experienced. Dainton thus rejects the traditional two-level view of experience, which emphasizes the act–content distinction. This is why his theory can be called "simple" or "one-level".

If there is no separate act of awareness that joins phenomenal contents together, then how come we do experience phenomenal contents together – both simultaneous contents and successive ones (e.g. C–D)? Dainton's model assumes that there exists a *basic relationship* that can join different contents. This is the relationship of the contents "being experienced together", or the contents being "co-conscious" for short. The idea is that such *co-consciousness* not only connects simultaneous experiences (synchronic co-consciousness), but that it also ranges a short way over time (diachronic co-consciousness).

The fact that co-consciousness extends over short periods implies that it creates temporally extended "phenomenal presents". To explain the continuity of our stream of consciousness, Dainton postulates that such successive phenomenal presents overlap.

It is important to note that the relationship of co-consciousness over time (diachronic co-consciousness) cannot be transitive. For example, consider the brief stream of consciousness consisting of the tones C–D–E. Suppose that the subject experiences C-being-followed-by-D, and D-being-followed-by-E. In saying this, Dainton assumes that the subject is directly aware of successions involving only two tones: by the time E occurs, C has dropped out of the subject's specious present. We thus see that C is co-conscious with D and D is co-conscious with E, but C and E are not co-conscious. Since the specious present is of limited duration (in Dainton's view probably less than a second), the relationship of co-consciousness over time (diachronic co-consciousness) cannot be transitive. Transitivity is restricted to the brief confines of the specious present.

Dainton now feels that he can both accommodate our direct experience of change and respect the phenomenal bonding constraint. It is the same relationship, namely direct co-consciousness, which links both experiences occurring at different times (e.g. C and D) and those occurring at the same time. He recognizes a problem, however. Even though co-consciousness is not transitive, it is symmetrical. Why then does our consciousness flow in a particular direction? If C and D are co-conscious, why do we hear C-being-followed-by-D rather than the other way round? Or, as he puts it, why does a persisting tone seem to continue on in a particular direction?

The only way Dainton can "account" for this feature is by presupposing or postulating it. He simply assumes that the contents that are symmetrically joined by co-consciousness must themselves possess an "inherent directional dynamism". A tone is not a "static auditory quality", but it is a "flowing quality". Dainton calls this feature "immanent flow" and says that it is an essential feature of any auditory content, as essential as timbre, pitch, or volume. The reason why we hear C as enduring for a brief while and then seamlessly running on into D is that these phenomenal contents possess such an "inherent and directed dynamic character". The idea is that the same applies to bodily sensations (e.g. pains and tickles), as well as to olfactory and gustatory contents. Not least, it applies in the visual case. He notes that when you see a bird fly you not only see the bird occupying a succession of different locations at different times, you see the bird *moving*. His suggestion is that this perceived movement is itself a dynamic and directed intrinsic feature of your visual experience, and just as essential to it as colour, size, or shape. He thus postulates that phenomenal contents quite generally possess the characteristic of immanent flow.

Dainton now thinks he is in a position to give a complete account of phenomenal temporality. Let us consider how he answers the questions we mentioned in Sect. 5.2 one by one.

- It seems obvious that we are directly aware of change, but how is this possible?
  Answer: It is possible because co-consciousness extends over time. This enables us to be aware of C-being-followed-by-D, for example.
- It seems obvious that we are directly aware of endurance, or phenomenal passage, but how is this possible?
  Answer: Phenomenal contents possess inherent dynamic character; they endure.
- It seems that the specious present is of short duration, but why would this be so?
  Answer: Because co-consciousness extends only a short way over time and is not transitive (at least in the diachronic case)
- It seems that consciousness flows in a particular direction, but why is this?

Answer: Because phenomenal contents possess not only an inherently dynamic but also an inherently directed character.

5.5 Problems with Dainton's View

Dainton has made a good job of presenting and criticizing various views about time consciousness, but his own view, though interesting in some respects, also seems to have some problems. I will first consider Revonsuo's recent criticisms of them, before adding some of my own.

5.5.1 Revonsuo's Critique

Revonsuo (2003) sees much to praise in Dainton's work, and we agree with that praise. However, Revonsuo also has criticisms:

> Dainton believes that to explain phenomenal unity and continuity, no reference to anything outside experience is required. Thus, he postulates a fundamental experiential relation called co-consciousness which is supposed to do all the explanatory work.

In contrast, Revonsuo thinks that to

> truly explain features of consciousness such as phenomenal unity and continuity, reference to mechanisms outside the phenomenal realm are necessary.

He continues:

> ... anyone who expects to find genuine explanations for the unity and continuity of consciousness will be disappointed. Co-consciousness is taken as the fundamental experiential relation that binds all the distinct phenomenal contents together and provides the unity and continuity of consciousness. But co-consciousness itself cannot be explained. Dainton's explanatory strategy forces him to stay within the phenomenal level as he constructs the explanations. I believe that many empirically minded scientists share my view that a genuine explanation goes beyond the level of description where the explanandum itself is identified. (Revonsuo 2003)

Presumably Revonsuo has in mind the idea that often a higher-level feature is explained in science by pointing to some lower-level features which make the higher-level feature necessary. For example, solidity as a macro-level property is explained by referring to factors such as intermolecular forces at the micro-level which make solidity seem necessary. By staying within the phenomenal level only, it seems Dainton cannot provide us with the kind of explanation of the unity and continuity of consciousness that would show how it necessarily follows from some other phenomena, such as the underlying neural phenomena. I agree with Revonsuo, and many cognitive neuroscientists, that it is certainly valuable to try to understand the phenomenal structure of consciousness in relation to other levels, such as the underlying neural processes. However, this does not have to mean that we reduce consciousness to the lower levels.

5.5.2 Further Criticisms of Dainton

Let us evaluate Dainton's model by examining some of its purported explanations one by one. Firstly, let us consider Dainton's suggestion that we are directly aware of change because co-consciousness extends over time. This feature enables us to be aware of C-being-followed-by-D, for example. But what is co-consciousness? Dainton does not tell us. The term refers to a "relationship" which we find in the phenomenal facts. Dainton calls this the basic relationship of "being experienced together". But surely a good account of time consciousness should tell us *why* certain contents are experienced together, not merely *that* they are experienced together. It seems that Dainton is merely giving a name to one aspect of the explanandum. If we accept that we hear C-being-followed-by-D, for example, we accept that C and D are experienced together. If we agree about what the phenomenon is, we can then proceed to try to make an explanation of the phenomenon. What needs explanation? Well, for example, we could ask *why* we experience C and D together. Of course, Dainton's notion of co-consciousness can be useful in the sense that it picks out one aspect of the explanandum, the fact that certain things are being experienced together. It is neutral as to whether the contents are simultaneous, or whether one is experienced slightly before the other. But still: conceptualizing the explanandum is not the same thing as giving an explanation. An explanation typically is an answer to a "why" question (see Faye (1999)).

Let us move on to further consider Dainton's account. He emphasizes that we are not only directly aware of change, but that we are also directly aware of endurance, or phenomenal passage. He suggests that such awareness of endurance is possible because phenomenal contents possess inherent dynamic character. This (together with his assumption that phenomenal contents are intrinsically conscious) might explain why we are aware of endurance. But why should phenomenal contents possess such inherent dynamical character, why should they endure? It seems he gives us no answer.

The next feature is the length of the specious present. Dainton assumes that the specious present is of short duration because co-consciousness extends only a short way over time and is not transitive (at least in the diachronic case). But if Dainton doesn't tell us why co-consciousness extends only a short way, or why it is not transitive, the explanation is quite empty. Again it is merely an attempt to describe the explanandum. Also, his estimate of the length (ca half a second) of the specious present may not be plausible in light of neuropsychology, for Pöppel's and Ruhnau's research suggests that the conscious "now" might be ca 3 seconds in duration (see Pöppel (1988) and Ruhnau (1995); Dainton (2000) acknowledges this possibility).

Finally, let us consider how Dainton explains the apparent fact that consciousness flows in a particular direction. He says this is so because phenomenal contents possess an inherently dynamic and directed character. Again, as long as he does not tell us why phenomenal contents have this kind of a character, the explanation seems very empty and ad hoc. For surely any good theory of time consciousness should explain – or at least make an educated guess – why consciousness flows or seems to flow in a particular direction.

Dainton's explanatory strategy seems dangerously close to one of just assuming or postulating something for everything that needs explanation. This is literally an ad hoc ("for this") strategy. In science, every explanation has to assume something, but the whole idea is that by paying the price for assuming certain things, one can then use those assumptions to explain a whole range of other things, far more than was initially assumed. Unfortunately, in Dainton's case it seems that he ends up postulating something for pretty much everything he is trying to account for. This is description, not explanation. Of course Dainton explicitly says that

> [m]y approach will be phenomenological: the task is the tightly circumscribed one of trying to formulate a *description* of our short-term experience of time that is clear, accurate and intelligible.

And he adds that he will ignore other interesting problems, such as how the brain manages to integrate the input from different sensory modalities to produce a real-time representation of the environment. But it also seems that this more humble approach of trying to provide a description has the tendency of being offered as an explanation.

Thus, Dainton talks about the need to provide "an adequate account of the temporal features of experience" (Dainton 2001, p. 94). Also, the various accounts he discusses, including his own, seem to be offered as explanations, not as mere descriptions.

In summary, Dainton may have offered us a useful way to conceptualize the explanandum, but it seems his account does not manage to explain phenomenal temporality. It leaves us with the following unanswered questions:

- Why should we assume that phenomenal contents are intrinsically conscious? What does such an assumption imply? In particular, does it imply that a certain type of dualism is true (in the sense that the world can be divided into intrinsically conscious phenomena and non-conscious phenomena, and that the distinction cannot be explained in any way?) Is it possible to explain the fact that phenomenal contents are conscious in some other way?
- Why is there co-consciousness, or why are certain contents experienced together?
- Why should we assume that co-consciousness extends only a short way and that it is not transitive?
- Why should we assume that phenomenal contents possess inherently dynamic character or that they endure? Is it possible to explain why they endure in some other way?
- Why should we assume that phenomenal contents possess an inherently directed character? Is it possible to explain their directedness in some other way?

Having seen that Dainton's model leaves many questions unanswered, let us move on to examine yet another approach which tries to explain our phenomenal temporality, namely Bohm's.

5.6 Bohm on Conscious Experience and Time

5.6.1 Bohm's Model of Phenomenal Temporality

As we saw in Chap. 3, David Bohm discussed the temporal structure of our conscious experience and its relation to underlying neural and physical processes in an interesting way in his *Wholeness and the Implicate Order* (Bohm 1980). Because his model is in some ways similar to the two-dimensional model, it will be useful first to remind ourselves of the latter and develop his model from that.

The basic idea of the two-dimensional model was that each momentary act consists of an awareness of a different momentary content (e.g. G) together with representations of the preceding contents (C–D–E–F-...). Past contents are presented under a distinctive temporal mode, degree of "presentedness". In a typical specious present the subject's span of direct awareness is then filled with contents of varying degrees of presentedness. The dynamic character of experience is explained by positing a two-dimensional time: there is the earlier–later ordering of acts of awareness, and the earlier–later ordering of contents presented to these acts. The specious present consists of a momentary awareness with a complex content that *seems* temporally extended. The appearance of a future-directed motion is a consequence of the ways in which the complex contents of different acts are interrelated; for example, tone C is represented in the contents of successive acts of awareness, but seems increasingly past in each. The model explains both how we can have an immediate experience of change and phenomenal flow (the latter being the impression we have that a qualitative unvarying sensation is continuously refreshed and renewed).

We remember that Dainton criticized the model for its phenomenal inaccuracy, the problem with the notion of presentedness, and for the fragmentary view of consciousness. Let us now develop Bohm's model and see whether it can answer Dainton's critique. Bohm's model is similar to the two-dimensional model in that he assumes that it is useful to think of consciousness in terms of a series of moments. The difference is that for Bohm such moments have a small extension; they can overlap. A moment typically has a certain explicit content which is a foreground and an implicit content which is a corresponding background. One moment gives rise to the next. In the latter moment the content that was implicate in the previous moment becomes explicate, and what was explicate in the previous moment becomes implicate. There can be a great deal of recurrence and stability in such a process, but there is also a possibility for a change.

When listening to music, for simplicity let's say our example melody: a rapid sequence of C–D–E–F–G-... played on the piano, Bohm would say that in the moment we hear G for the first time, we also hear C–D–E–F *reverberating* in consciousness. For Bohm these reverberations are not memories (contra memory-based accounts) nor representations (contra Broad's 2-D model) but rather *active transformations* of the original sounds. Instead of saying that past contents have varying degrees of "presentedness", Bohm would say they are all *co-present* but can differ in that they can have a different degree of *enfoldment*. The simplest way to measure the degree of enfoldment is to do this in terms of the time elapsed since the sound was first heard. Let us have as our unit of time 100 ms, and introduce an enfoldment or implication parameter n. C_n means that C was first heard n units ago, and is at the nth degree of enfoldment. If we now hear G for the first time, and have heard the sequence C–D–E–F–G with 100 ms intervals between the sounds, then we can describe the conscious experience of the melody at that moment as a co-presence of C_n , D_{n-1} , E_{n-2}, F_{n-3}, G_{n-4} (where $n = 4$). That is, we have *a co-presence of elements at different degrees of enfoldment.*

When listening to music, one is thus apprehending a set of co-present elements at different degrees of enfoldment. Such an experience has a certain order, which Bohm calls the enfolded or implicate order. We saw in Chap. 3 that when listening to music, one is, according to Bohm, directly experiencing an implicate order (as opposed to thinking about such an order abstractly in terms of thought). Bohm postulates that such a direct experience of the implicate order consists of an immediate sense of flow. This is not necessarily yet an intelligible explanation of why there is flow, but it does consider flow relationally rather than intrinsically. The idea is that our sense of flow is a certain mode of experiencing, namely a mode of directly experiencing an implicate order. Without the implicate order, there is no sense of flow. Listen to the set of notes so far apart in time that there is no reverberation, no co-presence of elements at different degrees of enfoldment, and the sense of flow we have when listening to music is gone. Listen to the set of notes closer together in time and flow is there. The same applies in vision, when watching a motion picture. There again one can see the different images that the eye picks up and sends to the brain as co-present elements at different degrees of enfoldment. And again, if the time interval between subsequent pictures is too great (more than 10th of a second), a sense of flow will disappear.

Bohm thus describes consciousness in terms of a series of moments. A given moment cannot be fixed exactly in relation to time (e.g. by the clock) but covers some vaguely defined and somewhat variable extended period of duration (Bohm 1980, p. 204). Each moment is experienced directly in the implicate order (and thus one expects to find a sense of flow whenever there are co-present elements at different degrees of enfoldment). Each moment of consciousness has a certain *explicit* content, which is a foreground, and an *implicit* content, which is a corresponding background. Bohm assumes that there is a certain "force of necessity" in the overall situation which has the effect that one moment of consciousness gives rise to the next moment. In such a later moment (part of the) content that was previously implicate can have become explicate while (part of) the previous explicate content can have become implicate.

Bohm's approach is thus in some important ways similar to the Husserlian/Broadian two-dimensional model. There is the earlier–later ordering of moments of consciousness, and the earlier–later ordering of contents apprehended in these moments. A new feature is the suggestion that the sense of flow arises because one is directly perceiving a set of co-present elements in different degrees of enfoldment.

In a typical moment of consciousness, the content of consciousness is thus a set of co-present elements which actively transform (e.g. musical tones) and are at different degrees of enfoldment. Let us then assume that subsequent moments of consciousness have a great deal of recurrence, so that they have partly the same content, in the sense of part of the active transformations being the same ones.

Let's say that moment A (which we now consider to be 500 ms long) involves hearing notes C, D, E, F, G, where C was heard first, and D, E, F, G followed at 100 ms intervals, in that order. To mark this, let us use the implication parameter n to mark the "degree of implication" or "degree of enfoldment" (basically the time that has passed since a tone was first heard). Let us further say that one unit of n is 100 ms. Thus, for C, at moment A, n would be 4. We can say that C has enfolded n times and write C_n, and for the other notes D_{n-1}, E_{n-2}, F_{n-3}, G_{n-4}.

Thus, we can describe a 500 ms moment of consciousness A by saying that it contains the following co-present elements at different degrees of enfoldment:

$$C_n$$

$$D_{n-1}$$

$$E_{n-2}$$

$$F_{n-3}$$

$$G_{n-4}$$

Let us say further that moment A gives rise to moment B, and that part of what was unconscious in moment A (say, the next tone that happens to be H) becomes part of the conscious content of B, while part of what was conscious in A (tone C) becomes part of the unconscious content of B. Tones D, E, F, G recur in moment B, though they all have a higher degree of enfoldment than in moment A. So, when the next moment of consciousness B unfolds, D, E, F, and G are still heard, but C has disappeared from consciousness, and H has just unfolded, like this:

$$D_n$$

$$E_{n-1}$$

$$F_{n-2}$$

$$G_{n-3}$$

$$H_{n-4}$$

But there is no reason to call the previously held notes "representations" of earlier experiences, for they are in some sense numerically the same as the previous (though their degree of enfoldment may have changed). This is a "process" ontology, but each tone constitutes a relatively autonomous element (having duration within a moment of consciousness and being recurrent from moment to moment), a phenomenal content which lives its life in conscious experience as an active transformation. We could call it an actively transforming phenomenal content.

In the above example, we have assumed that a moment of consciousness has a duration of 500 ms, but assuming that Ruhnau (2005) is correct, we could in some situations allow even 3 seconds duration for a moment of consciousness.

5.6.2 Bohm's Model and the Problems with the Two-Dimensional Model

Let us now see whether Bohm's model can answer Dainton's critique of the two-dimensional model. Dainton's first criticism was phenomenal inaccuracy. According to him, our consciousness is not clogged up with fading residues of prior experiences, and experiences do not always linger; they sometimes do end abruptly. If we cannot discern the posited representations, there is no reason to suppose that they have any phenomenological reality. The first point is that our experiences sometimes do linger (and even Dainton seems to imply this when he says they do not *always* linger). If that is a phenomenological fact, then any model that cannot do justice to it is phenomenologically inaccurate. Unfortunately, it seems that Dainton's own model is not very good at doing justice to lingering. It is thus phenomenologically inaccurate. What Bohm's model needs to explain is why it sometimes doesn't linger, but this shouldn't be too difficult. In Bohmian terms, the question is what determines whether elements (such as notes), whether unfolded or enfolded, are consciously experienced. This can depend on all kinds of factors and constraints, and we ought to be able to settle the issue, at least to some extent, by empirical means. This way we might understand why phenomenal contents can sometimes linger, while at other times they do not linger.

How about the problematic "presentedness" of Broad? This is no problem for Bohm, because the different active transformations are all present, while they have a different degree of enfoldment, which seems only natural.

Then there is the problem of fragmentation. Bohm's model postulates that consciousness can be described in terms of a series of moments. But these moments are not fragmented from each other. On the contrary, one moment gives rise to the next. Implicate content in the moment A becomes the explicate content of the subsequent moment B. This seems to be a phenomenologically accurate description, especially of our conscious experience of our thinking process. However, in a typical visual experience, the way moment A gives rise to moment B is not noticed by us. In this sense we may not get phenomenological evidence for Bohm's model. But Bohm's model is consistent with phenomenological data. Why should we (always) notice one moment giving rise to the next? Also, phenomenological experience need not accurately reflect what is "out there": it can "cheat us". For example, when we are watching a motion picture, "out there" is presented a set of discrete images very rapidly. But we experience a sense of flow (i.e. we are directly perceiving the implicate order in our brain/mind, or the set of co-present elements at different degrees of enfoldment). So, in this way our conscious experience sometimes gives us a "wrong" model of what is "actually" happening. The same could be going on when it comes to moments of consciousness themselves (as opposed to the phenomenal contents which are experienced in those moments). Consciousness may consist of discrete moments of consciousness, where a moment gives rise to the next. But instead of always experiencing this in a discrete way (a moment of consciousness A-giving-rise-to-B), we may just experience a continuity of consciousness. With conscious thought, when the whole process is slower, we may be better able to detect the way a moment gives rise to the next, and the way the implicate content of a moment A becomes the explicate content of a moment B, while what was explicate in A becomes implicate in B.

5.6.3 Bohm's Model and Dainton's Problems

Let us then see whether Bohm can answer the questions Dainton was originally concerned with.

- It seems obvious that we are directly aware of change, but how is this possible?
  Answer: For Bohm, our direct awareness of change (in the sense of movement) is a consequence of a direct perception of an implicate order in the brain, or a direct perception of a set of co-present elements at different degrees of enfoldment.
- It seems obvious that we are directly aware of endurace, or phenomenal passage, but how is this possible?
  Answer: Bohm postulates that there can be a great deal of recurrence and stability in subsequent, overlapping moments of consciousness.
- It seems that the specious present is of short duration, but why would this be so?
  Answer: The extent of the specious present can be variable, sometimes even 3 seconds (cf. Pöppel (1988) and Ruhnau (1995)). In the Bohmian scheme one might say that once the degree of enfoldment of an element becomes sufficiently high it disappears from consciousness. However, it can in principle re-unfold into conscious experience, depending on the circumstances.
- It seems that consciousness flows in a particular direction, but why is this?
  Answer: Often the direction of flow is given by the differences in the input to the brain. Why does the phenomenal car (in my phenomenal experience of watching a moving car) move from left to right? Well, trivially, because of the way in which the images picked up by the eye differ. Why is there a sense of flow? Because we are directly perceiving a set of co-present elements at different degrees of enfoldment. Putting these features together, we get a flow in a particular direction. Bohm has also tackled the problem of irreversibility in his research in physics, and it might be interesting to apply these ideas to the problem of the directedness of the flow of consciousness (see, for example, Bohm (1965, pp. 304–307).

Let us also consider how one would react from the implicate order point of view to the problems that Dainton's overlap seems to suffer from, namely:

- Why should we assume that phenomenal contents are intrinsically conscious? What does such an assumption imply? In particular, does it imply that a certain type of dualism is true (in the sense that the world can be divided into intrinsically conscious phenomena and non-conscious phenomena, and that the distinction cannot

be explained in any way?) Is it possible to explain the fact that phenomenal contents are conscious in some other way?
Answer: Bohm does not assume that phenomenal contents are intrinsically conscious. As we saw in Chap. 3, he assumes, exotically, that consciousness is enfolded in the holomovement. However, it seems that in the Bohmian scheme conscious awareness is something that only fairly complex living organisms have. We will further discuss the origin of consciousness in the Bohmian scheme in Chap. 6.

- Why is there co-consciousness, or why are certain contents experienced together?
Answer: The factor that explains why certain contents are experienced together is the overall implicate order which prevails in the brain/mind of the person in question. A conscious experience typically has an explicit content which is a foreground and an implicit content which is a corresponding background. The explicit content of a given moment typically unfolds from the implicit content of a previous moment. Bohm assumes that there is a force of "overall necessity" which brings certain enfolded elements in the brain/mind together to contribute to the explicit content that we most vividly experience in a given moment. Thus, this force of overall necessity accounts for why we experience certain explicit contents together. However, the implicit background is also a part of the overall conscious experience. The way the explicit foreground and the implicit background hang together (and can thus be experienced together) depends on the overall implicate order in the brain/mind.
- Why should we assume that co-consciousness extends only a short way and that it is not transitive?
Answer: Bohm might have been happy to allow that a moment of consciousness is longer than half-a-second, e.g. 3 seconds as Pöppel and Ruhnau suggest.
- Why should we assume that phenomenal contents possess inherently dynamic character or that they endure? Is it possible to explain why they endure in some other way?
Answer: Why does the sound C endure in Bohm's model? It "reverberates", it transforms actively within a single extended moment of consciousness. Of course, if the sound C is played continuously (e.g. by a cello), then this trivially explains why C endures in subsequent moments of consciousness. The reason why we experience contents enduring is that there is often a great deal of recurrence and sta-

bility in subsequent moments of consciousness, and the moments typically overlap..

- Why should we assume that phenomenal contents possess an inherently directed character? Is it possible to explain their directedness in some other way?
  Answer: I do not recall that Bohm explicitly discusses why phenomenal contents possess an inherently directed character. The directedness might be understood by examining the relations between the elements that are co-present. Altenatively, one could try to make use of Bohm's ideas on irreversibility in physics here (see, for example, Bohm (1965, p. 304–307).

5.6.4 Bohm in Relation to Revonsuo

Let us finally consider whether Bohm's scheme suffers from the criticism Revonsuo directs toward Dainton. In some ways, Revonsuo's criticism of Dainton is similar to the one we have presented above. Revonsuo thinks that to explain features of consciousness such as phenomenal unity and continuity, reference to mechanisms outside the phenomenal realm are necessary. Dainton offers co-consciousness as the explanation but it itself cannot be explained. The trouble is that Dainton stays within the phenomenal level as he constructs the explanations.

It seems clear that Bohm's explanations of conscious experience go beyond the phenomenal level, for he thinks that it is essential to consider conscious experience in relation to the underlying neural processes. Bohm's concept of implicate order helps us to think of the way the brain/mind processes information and the way that information is apprehended in a new way.

It is also interesting to compare and contrast Bohm's and Revonsuo's explanatory strategies in consciousness studies. To discuss this fully would require a chapter of its own, so short remarks have to suffice here. Revonsuo encourages us to approach conscious experience in terms of an explanatory strategy used in the biological sciences to explain a high-level phenomenon in terms of the lower-level features that could actually constitute and support the features of the phenomenon, and thus make it intelligible. This kind of "multi-level" mechanistic explanation has been very successful in the physical and biological sciences.

One thing to notice is that this explanatory strategy is in no way specifically "biological", for did it not originate from classical physics – from the molecular theory of heat and the kinetic theory of gases, where the feat was the (some say alleged) reduction of macro-level thermodynamics to micro-level statistical mechanics? Related to its roots in classical physics, this kind of multi-level biological explanation easily inherits the reductive spirit of classical physics, and leads to well-known problems when applied to conscious experience. In his writings, Revonsuo seems to imply that the higher-level feature, conscious experience, is thought to be literally "constituted of" some lower-level features (e.g. something like patterns of synchronized neural activity). But this easily denies conscious experience any autonomy and causal powers qua conscious, and raises the other traditional criticisms against brain–mind identity theory. For example, for many, the idea that conscious experience would be constituted by a pattern of (classical physical) neural activity seems to eliminate the experientality of conscious experience altogether.

Although Bohm's implicate order framework assumes that neuroscience has an important and indeed obvious explanatory role to play in consciousness studies, it is not committed to the kind of reductionism implicit in Revonsuo's multi-level approach (with its idea about the lower level always constituting the higher level). The implicate order framework originally arose from considering quantum and relativity physics, and is designed to handle radically holistic phenomena, where one can talk about a phenomenon (such as a quantum mechanical wave function) without the need to assume that the phenomenon is constituted of parts and their interactions at a lower level. It may be that an adequate scientific and philosophical discussion of the ontology of conscious experience requires a similarly holistic approach. Although conscious experience is intimately connected to and dependent on the underlying neural processes, it may be a mistake to try to reduce it constitutively to such processes (for an excellent and thorough presentation of Revonsuo's views, including those on the ontology of consciousness, see his 2005 book *Inner Presence*, especially Chap. 12.

In this chapter, we have demonstrated, albeit tentatively, that the implicate order framework allows us to develop a promising account of our experience of movement, both auditory and visual. I would like to conclude this chapter by emphasizing that although I have been critical of some of Dainton's views, I also think he has made a significant contribution to our understanding of time consciousness. His two recent books, *The Stream of Consciousness* and *Time and Space* are impressive achievements from which one can learn a great deal (see also Dainton (2004)). Indeed, his work – together with some of van Gelder's and Varela's recent views – provided the framework within which I began to consider and further develop Bohm's implicate order model of time consciousness. Similarly, Revonsuo's views have been very helpful to me in clarifying the questions involved and thus helping me to look for new perspectives.

6

Movement, Causation, and Consciousness

We have covered a lot of ground in the previous chapters, but many issues remain, of course, open for further discussion. In this final chapter I would like to touch briefly upon some of these. First of all, I will try to clarify Bohm's basic metaphysical proposal that "all is flux". Secondly, I will further consider the role of causality in Bohm's mind–matter scheme. Finally, I will discuss the prospects of solving the "hard problem" of consciousness in the Bohmian scheme.

6.1 Movement as Fundamental

We have seen in previous chapters how Bohm typically emphasizes that all our theories are limited. They are ways of looking at the world instead of literal descriptions of the way the world is. This applies especially to our metaphysical theories, or our views about the more general and fundamental nature of reality. They should not be seen as assumptions or conclusions about the final truth, but rather as proposals. Bohm requires of such proposals both a general self-consistency and consistency in the way they affect our life as a whole (Bohm 1980, p. 213). We have also seen that Bohm's basic metaphysical proposal is that "the totality of all that is" is *movement*. This is meant to replace the traditional metaphysical theory, according to which reality consists of some basic building blocks (particles and/or fields). He admits that such analysis worked reasonably well in physics until the end of the 19th century, but emphasizes that it does not fit with most of what has been discovered in 20th-century physics (Bohm 1976, p. 26). Of course, the analysis into parts can still be useful in a wide range of fields, but it should not be taken to be as fundamental as the proposal that reality is movement.

But how are we to make sense of the proposal that reality is more fundamentally movement? Of course, we have discussed this notion at various places in the previous chapters, but the reader may still be somewhat puzzled about what is meant by it. It might therefore be useful to try to further clarify what Bohm was driving at. In my view, he provides a particularly helpful explanation in a 1976 book *Fragmentation and Wholeness*, in a section entitled "On the Primacy of Movement and Undivided Wholeness, as Implied by Modern Physics" (Bohm (1976, pp. 36–41); republished in Nichol (2002)).

When we hear a proposal that "reality is fundamentally movement", this may still typically evoke an image of reality consisting of basic elements, particles and/or fields that move through space. In other words, "movement" seems to imply that there exists something (particle, field, etc.) that is "doing the moving". This is the assumption Bohm wants to challenge, emphasizing that movement is a much richer concept:

> ...movement means not just the motion of an object through space, but also much more subtle orders of change, development and evolution of every kind.(Bohm 1976, p. 40)

He gives a number of examples to illustrate this notion. Firstly, he considers what we might call "the movement of a symphony":

> This is expressed in terms of notes, which are sound waves (vibratory motions of molecules of air). But the essence of the movement cannot be understood in terms of such ideas of the motions of objects through space. (Bohm 1976, p. 40)

To bring out what he means, Bohm notes that it would be inappropriate to talk about the "motion of a symphony", except when, say, referring to the displacement of the orchestra through space on a train. He further notes that it would also be inappropriate to talk about the process of a symphony:

> Indeed the word "process" is based on the verb "to proceed" which means "to step forward". It thus refers to a particular kind of movement, which goes step by step, with one step following another. However, the movement of a symphony involves a total ordering, that is not essentially related to a process of time (though a process of time is involved in playing the notes, in a proper order). Indeed, one may in principle apprehend the whole movement of a symphony at any moment.
>
> It is clear then that in an art form such as music, one can have a direct sense of what it means for there to be a movement

> without some definite thing that is "doing the moving". (Bohm 1976, pp. 40–41)

With this example Bohm is seeking to establish a contrast between some usual, simple forms of movement (i.e. motion of an object through space, and a step-by-step process in time) and the more general notion of movement which involves "more subtle orders of change, development, and evolution of every kind". Given that he thinks that reality is movement in a more fundamental sense, the above suggests a metaphor that the universe is like the movement of a symphony. Thus, the essence of the universe is not the motion of objects (particles and/or fields) through space, nor a step-by-step evolution of the state of the universe in a process of time, although these notions of course can play some role. More fundamentally, the metaphor draws attention to the universe as a whole movement (indeed the "holomovement") in which a certain total ordering prevails. This total ordering is enfolded in each moment. The key point is that the total ordering of the movement that is the universe is not essentially related to a process of time. In this sense the holomovement is beyond time. It is the ground from which arise the sorts of processes in which a time order prevails.

Bohm next notes that the movement of a symphony is ultimately apprehended in a yet more subtle movement, namely the movement of attention. This is involved in

> all our sensory perceptions, and in the act of understanding the whole of perception and thought. (Bohm 1976, p. 41)

Attention is a psychological phenomenon which clearly seems to involve movement. Just as the movement of a symphony can be expressed in terms of vibratory motions of molecules of air, so the movement of attention might perhaps be expressed in terms of the motions of the particles and fields which constitute the various physiological processes in the "neural correlates" of attention (for example, involving thalamo-cortical loops, the extrastriate cortex, the anterior cingulate, etc. (see Baars et al. (2003, pp. 263–318)). But just as with the symphony, Bohm implies that the essence of the movement of attention cannot be understood in terms of such ideas of the motions of objects through space. Thus, the "motion of attention" might at best refer to the displacement of an attentive person through space on a train. Similarly, we do not talk about the "process of attention" in the sense that attention would be essentially a movement that goes step by step. Attention is thus not essentially a thing that moves in the brain, although attention involves movement of particles and fields, and various physiological processes in the brain.

Finally, Bohm considers yet another form of movement, namely the movement we find in a living being. This is the movement in which

> all the various functions are organized, to work together to create and maintain the whole organism... It is perhaps helpful here to note that the root of the word "to organize" is related to the Greek "ergon", which is based on a verb meaning "to work" and that this verb is also the root of the word "energy", which thus means literally "to work within". If we think of the movement of life as an "organizing energy" that is "working within" the movements in the organs, in the cells, and indeed, even in the atoms and elementary particles and thus ultimately merging with the universal field movement, this would perhaps help further in giving feeling for what it means to take movement as primary. (Bohm 1976, p. 41)

The above notion of "organizing energy" is, of course, very similar to the notion of active information introduced in Bohm and Hiley's ontological interpretation of quantum theory. At the quantum level, active information is contained in the wave function, and it organizes the movement of the particles.

6.2 Mental Causation

Thomas Nagel summarizes the mind–body problem neatly when he says that it seems obvious that materialism and dualism are incorrect, while it is not clear what the alternatives could be (Nagel 1995). In this book we have considered one such alternative, Bohm's non-materialistic and non-dualistic mind–matter proposal. Let us now consider whether this scheme allows us to make sense of one of the central aspects of the mind–body problem, namely the problem of mental causation. In brief, this is the problem about whether minds or mental properties can have any genuine influence upon the body and upon our actions. Do minds affect the body in virtue of their mental properties or in virtue of the physical properties that perhaps "instantiate" or "realize" the mental ones? Philosophers tend to express this question by asking whether minds are causally efficacious qua mental (where "qua" simply means "as")?

Of course, if one is a full-fledged materialist and assumes that minds are material, then the question of the causal influence of mind upon matter becomes a special case of causation in the material domain. That may have problems of its own (for example, the kinds that Hume drew attention to), but mind–matter causation would seem to be no more problematic than causation in general. However, many philosophers are unwilling to assume that minds are entirely material. As we have seen, the main reason for this is conscious experience and some of its properties, like qualia, intentionality, and the very fact that experiencing is going on. This then leads philosophers to postulate that mental properties are non-physical ones. However, this raises the problems of dualism. Where did these mental, non-physical properties come from? Can they have any genuine causal effect upon physical properties qua mental? If one assumes that they do, doesn't this create problems with physics, for example with the idea that the laws of physics are causally closed in the sense that non-physical things couldn't conceivably interfere with them? And if one assumes that mental properties qua mental have no causal influence at all upon the physical, what is the point of assuming their existence in the first place? As Jaegwon Kim has pointed out, an existence without causal powers is an existence hardly worth having (Kim 1989).

The above brings out some of the reasons why it seems that both materialism and dualism are likely to be incorrect. Let us now consider how Bohm's mind–matter proposal might tackle the issue of mental causation. As opposed to materialism, which assumes that matter is the ground of existence, Bohm assumes that the ground of existence is simply movement in which a multi-dimensional implicate order prevails. In effect, this is also to deny dualism, as, strictly speaking, only one "substance", namely movement, is assumed. But what is this "movement"? Is it mental or physical, or perhaps both? When extending the implicate order scheme with the ontological interpretation, we saw that Bohm and Hiley introduced the notion of active information as fundamental. Because the information is essentially active, it is a kind of movement. For example, the active information at the quantum level involves the movement of the subtle quantum field, a movement which gives rise to particles (in quantum field theory) and guides the movement of particles (in quantum particle theory). Bohm assumes further that active information always has both a mental and a physical aspect. Thus, the mental and the physical are assumed to be aspects of a single underlying reality, instead of separate substances in interaction.

Of course, one might say that information is a more mental than physical concept, when considered in terms of the traditional notions of mind and matter. Thus, Bohm's proposal that the essence of reality is active information could be seen as a form of objective idealism (see, for example, Niiniluoto (2000)). However, Bohm was keen to emphasize that even the more subtle kinds of information (e.g. that contained in conscious experience) always have a physical aspect. This means that such information can, in principle, influence the course of physical processes without violating the principle of the causal closure of the physical. Perhaps the correct way of characterizing Bohm's monism in terms of traditional categories is to say that it is more on the mental side in the matter–mind scale, but not, strictly speaking, a form of objective idealism. I also recall that this is the way Bohm himself characterized his viewpoint when we discussed the issue in the late 1980s and early 1990s (see also Bohm (1990, p. 285) for his rejection of "idealism").

How do we then understand the two-way traffic between mind and body in Bohm's ontology? If we consider a given level, then the key idea is that active information, being simultaneously both mental and physical, bridges the mental and physical sides. But we also saw that Bohm introduces the notion of a hierarchy of levels. The more subtle levels correspond to what we traditionally call "mind" and the less subtle ones to what we call "body". The more subtle levels gather information about the less subtle ones via a soma-significant process. Once the meaning of this information is apprehended, there is a reverse, signa-somatic process in which the more subtle, "mental" levels organize the less subtle, "physical" ones. If we call the more subtle levels "mental" and the less subtle "physical", then there is a sense in which the mental qua mental affects the physical, and vice versa.

However, there is also another sense in which we can ask in the Bohmian scheme whether the mental affects the physical. For each level of active information has both a mental aspect and a physical aspect. Can we say there that the mental aspect, qua mental, affects the physical aspect, and vice versa? Consider the "prototype" case of the electron in the ontological interpretation of quantum particle mechanics. The electron has two aspects, the field aspect and the particle aspect. The field contains active information, which guides the behavior of the particle aspect. There are also some suggestions that the particle could affect its field (Jack Sarfatti has called this "back-action" (Sarfatti 1997)). In this context we could say that the field aspect is "(proto)mental" and the particle aspect "physical". Thus, even the behavior of an electron involves what we might call "proto-mental causation". The information is carried by the subtle quantum field, and it cannot be reduced to mechanical components. Thus, if we look at an individual electron, it has an irreducible proto-mental aspect, the quantum field, which has an influence upon its physical aspect, the particle. If a theory of back-action can be worked out coherently, then we also have a prototype model of the way a physical aspect affects a proto-mental aspect, thus opening up the way to discuss a two-way traffic between the physical and the mental.

There is an ambiguity concerning causality in the Bohmian mind–matter theory (cf. Guarini (2003)). In the early implicate order framework, Bohm suggests that the relationship of mind and body is a non-causal one. "Non-causal" here means that mind and body are correlated projections of an underlying common ground, rather than separate substances in interaction. However, in the extended implicate order framework, it is also possible, as we have seen, to discuss a causal two-way traffic between the more subtle (mental) and less subtle (physical) levels. These "levels" should not, however, be taken to stand for separate parts of reality, but are convenient abstractions we make from the undivided totality. Nor should the relationship between the levels be seen as a mechanical interaction in the spirit of classical physics (except perhaps at the most manifest levels where the laws of classical physics provide a good approximation). In general, implicate order is the relationship that prevails between the levels. The more subtle levels enfold or gather information about the less subtle ones, and in turn act to organize the latter on the basis of the meaning of the information thus gathered. This organizing activity can be seen as a kind of unfoldment of the meaning of the enfolded information.

What then is the role of non-locality and the idea of mind and matter as non-causally connected, correlated projections in the extended implicate order framework? It seems that Bohm thinks that the levels are, in the end, also projected from the multidimensional implicate order. There is the domain of the sort of matter which is approximately describable in terms of classical physics. This domain is part of what Bohm and Hiley call the "total quantum world", but it is a part in which the quantum potential has a negligible effect (Bohm & Hiley 1993, p. 179). However, there is also the domain of the physical world in which the quantum potential has a significant effect. This can be called the subtle part of the quantum world (Bohm & Hiley 1993, p. 179). At this level, non-locality prevails. Bohm and Hiley speculate that such non-locality might also play a role in the functioning of the brain:

> ... the brain would be a system that can, like a measuring apparatus, manifest and reveal aspects of the quantum world in the overall processes ... as the process of perception unfolds into the

> brain, it may as it were connect to the subtle quantum domain which latter may in turn reconnect to the classical domain, as outgoing action is determined through amplification of quantum effects.
>
> Moreover, it does not necessarily follow that quantum effects will only be of importance in the domain of very small energies. For ... it is just through certain kinds of non-locality that locality can emerge, for example in a measuring apparatus. Similar non-locality may be required for the brain to have a local and essentially classical sub-domain of function. (Bohm & Hiley 1993, p. 179)

The above provides one way of thinking of the causal traffic between mind and matter in the framework in which the implicate order is extended with the ontological interpretation. The information gathered in perception may, in a soma-significant process, connect to the more subtle quantum domain where its meaning is apprehended. This meaning then acts "downwards" so that quantum effects are amplified. Of course, we have seen that Bohm proposed that the quantum field is just one part of the overall process, and that it is likely that mental processes and consciousness involve more subtle levels. However, the quantum field is important because it provides one prototypical suggestion, and even a mathematical description, about what the nature of the "subtle levels" might be, and how they connect with the less subtle, "manifest" levels.

Bohm and Hiley also emphasize above that certain kinds of non-locality are necessary for the emergence of locality, including, perhaps, even the local functions in the brain (i.e. functions which can be described approximately in terms of classical physics). Thus it seems that non-locality is fundamental even in the extended implicate order framework, and indeed the feature that makes locality possible. If we say that non-locality is a non-causal relationship, while locality fits well with our usual ideas of causality, it follows that it is non-causality which makes causality possible. Thus, although it is reasonable to apply the notion of local causality both in some domains of the physical world, and to describe a part of the mind–matter relation, it is the non-local and non-causal relationship that seems to be the fundamental one in the Bohmian scheme. Leibniz might have approved.[1]

[1] For a further discussion of the problem of mental causation in the light of the ontological interpretation of quantum theory, see Pylkkänen (1992 & 1995) and Hiley and Pylkkänen (2005).

6.3 How is an Experiencing Physical System Possible?

In this final section of the book we will discuss the hard problem of consciousness in the light of the Bohmian scheme. In the emerging discipline of consciousness studies, perhaps more so than in other developing disciplines, the formulation of the problems to be studied has played a major role. Given that so much seems obvious about at least human consciousness, there is much to study that doesn't differ from other scientific areas of study. One can, for example, try to find regularities in the way our degree of being conscious varies in the sleep–wake cycle, and then measure the state of the underlying neural processes and see whether regularities there match with the experienced regularities. It is possible to study the different sensory modalities in a similar way. All this is straightforward in principle; there are problems that can be clearly formulated and we know what to do to address them.

However, as is widely acknowledged, there is also something about consciousness that is puzzling or problematic in an entirely different way. It is particularly puzzling if one – like many contemporary researchers – is a physicalist or a materialist about the mind and believes that consciousness should be explainable in terms of purely physical or neurobiological terms, in some sense derivable from physical and neurophysiological processes, or perhaps from computation implemented in such processes. Now, it is not at all obvious *why* a purely physical or neurophysiological process, however complex a computation it may be implementing, should have consciousness associated with it. This is, of course, what Leibniz suggested with his analogy of the mill, which we referred to in Chap. 1.

We can put the above in another way: it seems *obvious* that humans and animals are conscious, but it is not at all obvious *how* conscious experience could arise from their physical and neurobiological organization. This latter is, roughly, what the philosopher David Chalmers (1995, 1996) has called the "hard problem" of consciousness, as opposed to what he calls the "easy" problems to do with explaining what happens in consciousness (e.g. unification) in terms of underlying neurobiological structures and functions (e.g. neural synchrony).

The hard problem is a general problem: why is there *any* experience associated with *any* physical process *ever*? The problem has to do with our assumptions and intuitions about the general, fundamental nature of systems and processes. It seems that widely held scientific and philosophical intuitions about physical and biological systems are such that it is completely puzzling why a physical/biological system ought to or even could have conscious experience as an aspect of the system. And yet at the same time even more widely held everyday intuitions say that it is obvious that certain biological systems, namely human beings and at least some animals *are* conscious. *It is obvious that we are conscious but it is puzzling how we can be conscious, given our intuitions about physical systems.*

The situation can be characterized by saying that it seems obvious that human beings and at least certain animals are instances of what we can call an *experiencing physical system*. This we can take as given. The question then becomes: *How is an experiencing system possible?* The philosophical reader will recognize the Kantian spirit of the "methodology" here. Kant took it to be a fact that there are synthetic a priori judgements, and a major aim of the *Critique of Pure Reason* was to show *how* such judgements are possible. Similarly the challenge of contemporary consciousness studies is to show how an experiencing physical system is possible *at all* (cf. Bieri (1995)).

The hard problem debate suggests that it is very difficult to give a naturalistic explanation of consciousness in the prevalent general framework of contemporary natural science, especially biology and neuroscience. Now, one feature of this framework is that it is *mechanistic*. At the same time, as we have indeed emphasized throughout this book, contemporary physics implies that a different framework of concepts is required to describe nature at a general level, a framework which we can characterize as *holistic*, to contrast it with mechanism. My general conjecture in this book has been that it will turn out to be *easier* to give a naturalistic explanation of consciousness in the new holistic framework that has been emerging from contemporary physics, especially in the Bohmian scheme. In a similar vein, I suggest that the failure of contemporary neuroscientists and philosophers of mind to come to terms with consciousness is partly due to a tacit *overcommitment* to the mechanistic framework. The mechanistic framework has been very successful in physics, biology, and even cognitive science, but it seems obviously limited to describing the fundamental physical phenomena. As we have emphasized, it worked well in physics until the end of the 19th century, but has pretty much failed to fit with new major developments in physics after that.[2]

[2] Of course, it is possible to broaden the meaning of "mechanical" beyond "push–pull" systems and accommodate at least some aspects of quantum processes under a broader notion of a "mechanism". It is important, however, to realize just how radically quantum theory violates the traditional notion of mechanism. For an interesting contemporary discussion of mechanisms in terms of entities and activities, see P. Machamer et al. (2000).

This is an important point. In my experience, many psychologists, philosophers, and neuroscientists seek a mechanistic explanation of consciousness, because they assume – perhaps tacitly – that scientific explanations are always, as a matter of principle, mechanistic. This is why it is important to realize that mechanistic explanations do not really work to describe the more fundamental levels of the physical world. Now, if there is one domain of nature where mechanistic explanations do not work, this makes it possible, in principle, that there can be others. It is my conjecture that conscious experience has aspects that defy mechanistic explanations. In this sense, conscious experience is similar to, say, quantum phenomena. Of course, we have also seen in this book that Bohm has suggested that there is a more profound similarity between quantum phenomena and conscious experience. For he proposed that both are essentially movement in which the implicate order prevails. In this sense, both conscious experience and quantum phenomena have strongly non-mechanistic aspects.

In my view, the limits of the mechanistic approach in the description of consciousness have also been brought out well by the Australian philosopher David Chalmers, for example in his 1996 book *The Conscious Mind*, and I will now move on to discuss Chalmers' views, before relating them to Bohm's views. A brief comparison of their views is useful in order to see how the Bohmian scheme might deal with the hard problem of consciousness.[3]

6.3.1 David Chalmers' Approach

David Chalmers not only gave a particularly succinct and influential formulation of the hard problem of consciousness, he also proceeded to offer his suggestion about how to approach it. Chalmers' views have been widely and thoroughly presented and discussed in the literature, so a brief presentation will suffice here. He mentioned four premises which, roughly, provide a ground for his view (Chalmers 1996, p. 61):

1. Conscious experience exists.
2. Conscious experience is not logically supervenient on the physical.
3. If there are facts that are not logically supervenient on the physical facts, then materialism is false.
4. The physical domain is causally closed.

[3] For a preliminary discussion of the relationship between Bohm's and Chalmers' views about information, see also Pylkkänen (2000).

Chalmers' view, which he calls "naturalistic dualism", is, roughly, an inference from the above premises. On the basis of thought experiments, he concludes that physical facts do not necessitate phenomenal facts. A useful thought experiment here is a famous one presented by Kripke. Suppose that the world does not yet exist and that God creates our world by first creating physical facts. The question is: does God have to do extra work to create phenomenal facts? According to Chalmers, extra work is required, as phenomenal facts do not come for free with physical facts. More technically, phenomenal facts are not logically contained in physical facts, and they are not logically supervenient or dependent on them.

Now, Chalmers takes it as given that phenomenal facts exist. If there are facts which cannot be derived from the known fundamental facts of the universe (e.g. the fundamental physical properties), these facts might well themselves be fundamental. There are then two types of fundamental facts, physical and phenomenal ones, and a kind of "naturalistic dualism" emerges. The challenge, he says, is then to do what physics has successfully done, namely to connect fundamental properties with the help of fundamental laws. When it comes to phenomenal properties, the laws in question must be fundamental psychophysical laws, relating phenomenal properties with certain physical properties, and in this way explaining the observed and obvious co-occurence between physical and phenomenal properties that we see in ordinary experience.

Chalmers' task is then to search for the psychophysical law. He begins with high-level principles such as the idea that *consciousness correlates with awareness*. Note that awareness is for him a functional notion, which tries to capture our access to some information and ability to use that information in the control of behavior. Awareness does not refer to phenomenal aspects.[4] The principle says that where there

[4] Chalmers uses the term awareness to denote a psychological property associated with experience itself or with phenomenal consciousness. He distinguishes between the phenomenal and the psychological concepts of mind. The phenomenal one is the concept of mind as conscious experience. The psychological one, common in cognitive science, is the concept of mind as the causal or explanatory basis of behaviour. According to the psychological concept, it matters little whether a mental state has a conscious quality or not. What matters is the role it plays in a cognitive economy. Thus "[a]wareness can be broadly analyzed as a state wherein we have access to some information, and can use that information in the control of behaviour. One can be aware of an object in the environment, of a state of one's body, or of one's mental state... Awareness of information generally brings with it the ability to knowingly direct behaviour depending on that information. This is clearly a functional notion." (Chalmers 1996, p. 28). To explain awareness,

is consciousness, there is awareness, and where there is (the right kind of) awareness, there is consciousness (Chalmers 1996, p. 223). There is also the principle of *structural coherence*, which says that the structure of consciousness is mirrored by the structure of awareness, and the structure of awareness is mirrored by the structure of consciousness. Finally, there is the principle of *organizational invariance*, which holds that given any system that has conscious experiences, then any system that has the same fine-grained functional organization will have qualitatively identical experiences.

Such high-level principles are useful but do not have the kind of status of fundamental laws which Chalmers thinks is required to connect the fundamental properties of the universe, the physical and the phenomenal. As a candidate for a more fundamental law he proposes a double-aspect theory of *information*. In a typical conscious experience like that of colour, it seems that the same information is embedded both in the neural process and in phenomenal experience. Is it not reasonable to speculate that information there has two sorts of properties, the phenomenal ones and the physical ones? And if information has phenomenal properties in human conscious experience, perhaps it has them in a more general sense? These kinds of considerations lead Chalmers to propose that information is a fundamental feature of the world which has both physical and phenomenal properties as fundamental properties.

Chalmers seeks to ground the view that information is fundamental upon contemporary physics. He considers the so-called "it from bit" view advocated by Wheeler, according to which physics implies that information is fundamental to the physics of the universe, and even that physical properties and laws may be derivative from informational properties and laws. He understands the "it from bit" view as providing us with

> a picture of the world as pure informational flux, without any further substance to it. (Some versions of the view may also allow space-time as a primitive framework within which the information spaces are embedded; other versions see space-time itself as constituted by the relations among information spaces.) The world is simply a world of primitive differences, and of causal and dynamic relations among those differences. On this view,

it is enough to explain the physical and computational functioning of a system. I must confess that I find this non-conscious or "a-conscious" notion of awareness not very helpful.

> to try to say anything further about the world is a mistake. (Chalmers 1996, p. 303)

Chalmers mentions two problems with this view of the world. Firstly, phenomenal properties have an intrinsic nature, one that is not exhausted by their location in an information space, and it seems that a purely informational view of the world leaves no room for these intrinsic qualities. Secondly, the world may seem too lacking in substance to be a world; for according to this view, the world consists of causal relations which relate information states, but these states are remarkably insubstantial, being merely different from each other and having no nature of their own. He notes that one might find this picture of a world without intrinsic nature not to be a picture of a world at all.

He then tries to resolve both problems together: perhaps the intrinsic nature required to ground the information states is closely related to the intrinsic nature present in phenomenology. Perhaps, he says, one is even constitutive of the other.

> The suggestion is that the information spaces required by physics are themselves grounded in phenomenal or protophenomenal properties... Physics requires information states but cares only about their relation, not their intrinsic nature; phenomenology requires information states but cares only about the intrinsic nature. This view postulates a single basic set of information states unifying the two. We might say that internal aspects of these states are phenomenal, and the external aspects are physical. Or as a slogan: Experience is information from the inside; physics is information from the outside. (Chalmers 1996, p. 305)

This view of Chalmers has interesting similarities with Bohm's notion of active information. There are, however, differences. Chalmers is more explicit about phenomenal aspects, which makes his notion of information potentially a more powerful tool for the description of consciousness. At the same time, Bohm's view seems grounded in more detail upon quantum theory, and is thus a more concrete and physically plausible proposal about the place of information in nature. Let us thus consider the relation of these two schemes.

6.3.2 Bohm vs. Chalmers on the Hard Problem of Consciousness

It seems natural to ask whether it would be possible to combine the strong aspects of the Bohmian and Chalmersian schemes and that way to arrive at a more comprehensive theory. It is important to keep in mind, however, that the notions of information Bohm and Chalmers respectively use are not the same. Bohm's notion of active information is not essentially related to Shannon information, which is expressed in terms of bits (see Bohm and Hiley (1993); Maroney (2002)). In contrast, the notion of information Chalmers refers to is closer to Shannon information. However, there is likely to be a deeper unity behind these notions, and thus it seems reasonable to try to combine the perspectives, as long as one is aware of the differences.

There are at least two problems with Chalmers' view that have been discussed in the literature. First of all, there is the problem of epiphenomenalism, the causal inefficacy of the mental, which we discussed earlier in this chapter. Chalmers gives high value to the principle that the physical domain is causally closed in the sense that the fundamental phenomenal properties, being non-physical, cannot influence any physical event. This makes the phenomenal causally inefficacious, which in turns raises questions about its very existence (see, for example, Kim (1989)).[5]

Secondly, there is the problem that arises from the fact that information seems ubiquitous. For if we first say that information is ubiquitous and then say that information always has, as a matter of a fundamental law, phenomenal properties, this seems to make phenomenal properties ubiquitous. Thermostats, for example, embody information; are they therefore conscious? Chalmers proposes that we either constrain the fundamental laws so that there is experience associated with only very high-level information processing or then bite the bullet and allow simple experience where there is simple information processing, such as with thermostats. He writes:

> My own intuition is that there may well be a constraint on the double-aspect principle but information in simple systems such as thermostats might qualify all the same. (Chalmers 1996, p. 301)

---

[5] In discussions that took place in the University of Skövde workshop "Can there be a science of consciousness?" in June 2000, Chalmers said that he does not hold rigidly to the principle of the causal closure of the physical domain. He is willing, for example, to consider giving it up in the context of Stapp's interpretation of the quantum theory, which has some appeal to him. In his earlier published work he seems, however, to favour the option of saving the idea of the physical domain as a closed causal network and assuming that experience inheres in the otherwise empty nodes of this network, by virtue of a Russellian monism on which the intrinsic properties of matter are proto-experiental (Chalmers, 1997, p. 45).

It seems to me that Bohm's notion of active information can tackle better both the problem of epiphenomenalism and the problem of the ubiquity of phenomenal properties. Regarding epiphenomenalism, Bohm's notion of soma-significance implies that the content of the information at a more subtle level can have an effect upon the physical processes at more manifest levels. If we assume (somewhat similarly to Chalmers) that phenomenal properties are a kind of content of the active information (at least when the active information is in the brain/mind), then the Bohmian scheme suggests that phenomenal properties can influence physical properties.

Regarding the problem of the ubiquity of phenomenal properties, it seems Bohm was unwilling to postulate that, say, electrons have phenomenal properties. He thought it obvious that electrons are not conscious, and presumably this also implies that they do not have phenomenal properties. In this way, Bohm perhaps avoids the problem of the ubiquity of phenomenal properties which arises in Chalmers' scheme. However, Bohm is then left with the problem of explaining why there are phenomenal properties and conscious experience associated with the brain/mind. Why does active information in the brain/mind have phenomenal properties, while active information associated with a single electron does not? In the Bohmian scheme, one might approach this by saying that conscious experience comes in at the more "subtle levels", but, of course, much more needs to be said before such a qualitative hunch becomes a good theory.

Bohm did not directly tackle the hard problem of consciousness in his research, and it is therefore somewhat difficult to say how he would have approached it. It seems clear, however, that he did not deny the reality of conscious experience. On the contrary, it seems he thought that consciousness is a fundamental aspect of reality, an aspect that is enfolded in the holomovement, and which is projected from there in certain conditions, along with the body with which it is correlated. In this sense, Bohm seems to have assumed that consciousness is a fundamental feature of the world, at least as fundamental as matter. At the same time, the implicate order restricts consciousness in the sense that only systems in the explicate order which are sufficiently complex or subtle are conscious. If you like, the fact that the implicate order prevails in the holomovement is responsible for the fact that human beings and higher animals – as relatively autonomous subtotalities – are conscious, while electrons are not. The origin of consciousness in the Bohmian scheme is likely to be in the "depths" of the implicate order rather than in the interactions of mechanical elements in the explicate order.

One way to express the hard problem of consciousness in Bohmian terms is to say that there seems to be nothing within the explicate order that would necessitate or make possible conscious experience. Traditional philosophy of mind and neuroscience often assume that the explicate order is all there is to the physical world, while at the same time seeking to locate consciousness to the physical world. Thus it is not surprising, from the Bohmian point of view, that the hard problem is so acute for these subjects. The key point is that the Bohmian scheme proposes that there is more to the world than the explicate order of matter, namely the implicate order and what may be beyond that. If conscious experience cannot be fully accounted for within the explicate order of matter, then there is no choice but to explore the role played by the implicate order and what may lie beyond it.

Let me conclude by very briefly sketching how one could begin to explore the origin of consciousness in the Bohmian scheme. It seems that in this scheme, conscious experience involves the interplay between explicate and implicate orders. The content of conscious experience that we are most easily aware of is the explicate order, most vivid in our visual experience of a static scene, such as the furniture in a room. However, by just looking at this explicate order (including its neural correlate in the brain/body) we will not find an explanation for why it is conscious. Now, an explicate order typically unfolds from an implicate order. For example, the explicate order of my visual experience unfolds from the overall implicate order that prevails in the various levels of the brain/mind. One possibility is that the unfolded, explicate order is somehow apprehended or surveyed by some more subtle levels of activity at more implicate levels, and that conscious experience arises from such apprehending. This idea is somewhat similar to the so called "higher-order thought" and "higher-order perception" theories of consciousness (see, for example, Carruthers (2001)). However, perhaps the new notion of subtle, implicate levels in the brain/mind gives the Bohmian scheme more explanatory power than is enjoyed by the usual higher-order theories.

In the traditional materialistic scheme, consciousness is an anomaly, a mystery in a mechanical universe. In Bohm's new scheme, which is based on quantum and relativity physics, consciousness exhibits the same implicate order which prevails in both inanimate and animate matter. The Bohmian universe is thus more "consciousness-friendly" than the universe of classical physics and contemporary neuroscience, which are typically mechanistic. However, Bohm's scheme in its current state does not answer all the puzzling questions about consciousness that have been raised in the contemporary debate, such as the hard problem of consciousness. But perhaps it provides one framework in which we may hope to develop better theories in the future.

Bibliography

Albert, D. Z. (1992): *Quantum Mechanics and Experience* (Harvard University Press, Cambridge MA)

Albert, D. Z. (1994): Bohm's Alternative to Quantum Theory, *Scientific American*, May 1994

Aspect, A., Dalibard, J., & Roger, G. (1982): Experimental Test of Bell's Inequalities Using Time-Varying Analyzers, *Phys. Rev. Lett.* **49**, 1804–1807

Atmanspacher, H. (2004): Quantum Approaches to Consciousness, in *The Stanford Encyclopedia of Philosophy*, Winter 2004 edn, ed. by E. N. Zalta (http://plato. stanford.edu/archives/win2004/entries/qt-consciousness/)

Atmanspacher, H. & Ruhnau, E. (eds) (1997): *Time, Temporality, Now. Experiencing Time and Concepts of Time in an Interdisciplinary Perspective* (Springer, Berlin, Heidelberg)

Audi, R. (1998): *Epistemology: A Contemporary Introduction to the Theory of Knowledge* (Routledge, London)

Baars, B. (1988): *A Cognitive Theory of Consciousness* (Cambridge University Press, Cambridge)

Bacciagaluppi, G. (2005): The Role of Decoherence in Quantum Mechanics, in *The Stanford Encyclopedia of Philosophy*, Summer 2004 edn, ed. by E. N. Zalta (http://plato.stanford.edu/archives/sum2005/entries/qm-decoherence/)

Bedard, K. (1999): Material Objects in Bohm's Interpretation, *Philos. Sci.* **66**(2), 221–242

Bekenstein, J. D. (2003): Information in the Holographic Universe, *Scientific American*, August 2003

Bell, J. S. (1987): *Speakable and Unspeakable in Quantum Mechanics: Collected Papers on Quantum Philosophy* (Cambridge University Press, Cambridge)

Beller, M. (1999): *Quantum Dialogue: The Making of a Revolution* (University of Chicago Press, Chicago)

Bergström, M. (1969): An Entropy Model of the Developing Brain, *Devel. Psych.*, **2**(3), 139–152

Bieri, P (1995): Why is Consciousness Puzzling?, in *Conscious Experience*, ed. by T. Metzinger (Imprint Academic, Schöningh), pp. 45–60

Block, N. (1978): Troubles with Functionalism, in *Minnesota Studies in Philosophy of Science*, vol. IX, ed. by C. V. Savage (University of Minnesota Press, Minneapolis)

Block, N. (1980): Introduction: What Is Functionalism?, in *Readings in Philosophy of Psychology* vol. 1, ed. by N. Block (Harvard University Press, Cambridge MA)

Block, N. (1990): Can the Mind Change the World?, in *Meaning and Method: Essays in Honor of Hilary Putnam*, ed. by G. Boolos (Cambridge University Press, Cambridge)

Block, N., Flanagan, O. J., Güzeldere, G. (1997): *The Nature of Consciousness: Philosophical Debates* (MIT Press, Cambridge MA)

Bogen, J. B. (1995): Relations, Internal and External, in *The Oxford Companion to Philosophy*, ed. by T. Honderich (Oxford University Press, Oxford)

Bohm, D. (1951): *Quantum Theory* (Prentice-Hall, New York) [also Dover, New York, 1989]

Bohm, D. (1952): A Suggested Interpretation of the Quantum Theory in Terms of Hidden Variables I & II, *Phys. Rev.* **85**, 166–189

Bohm, D. (1957): *Causality and Chance in Modern Physics* (Routledge, London) [2nd edn with new preface, 1984]

Bohm, D. (1965a): *The Special Theory of Relativity* (Routledge, London)

Bohm, D. (1965b): Problems in the Basic Concepts of Physics, in *Satyendranath Bose 70th Birthday Commemoration Volume*, part II (Prof. S. N. Bose 70th Birthday Celebration Committee, Calcutta) [an inaugural lecture delivered at Birkbeck College, University of London, 13 February 1963, with two appendices]

Bohm, D. (1965c): Space, Time, and the Quantum Theory Understood in Terms of Discrete Structural Process, in *Proceedings of the International Conference on Elementary Particles 1965, Kyoto*, pp. 252–287

Bohm, D. (1965d): A Proposed Topological Formulation of the Quantum Theory, in *The Scientist Speculates*, ed. by I. J. Good (Putnam, New York)

Bohm, D. (1969a): Some Remarks on the Notion of Order, in *Towards a Theoretical Biology, 2. Sketches*, ed. by C. H. Waddington (Aldine, Chicago)

Bohm, D. (1969b): Further Remarks on Order, in *Towards a Theoretical Biology, 2. Sketches*, ed. by C. H. Waddington (Aldine, Chicago)

Bohm, D. (1970): Fragmentation in Science and in Society, *Impact Sci. Soc.*, **XX**(2), 159–169

Bohm, D. (1971): Space-Time Geometry as an Abstraction from "Spinor" Ordering, in *Perspectives in Quantum Theory: Essays in Honor of Alfred Landé*, ed. by W. Yougrau & A. van der Merwe (MIT Press, Cambridge MA)

Bohm, D. (1973): Human Nature as the Product of our Mental Models, in *The Limits of Human Nature*, ed. by J. Benthall (Allen Lane, London)

Bohm, D. (1974) On the Subjectivity and Objectivity of Knowledge, in *Beyond Chance and Necessity: A Critical Inquiry into Professor Jacques Monod's Chance and Necessity*, ed. by John Lewis (Humanities Press, Atlantic Highlands NJ)

Bohm, D. (1976): *Fragmentation and Wholeness* (The Van Leer Jerusalem Foundation, Jerusalem) [republished in Nichol (2002)]

Bohm, D. (1977): Science as Perception-Communication, in *The Structure of Scientific Theories*, ed. by F. Suppe (University of Illinois Press, Chicago)

Bohm, D. (1979): David Joseph Bohm – Interview, in *A Question of Physics: Conversations in Physics and Biology*, ed. by P. Buckley & F. D. Peat (Routledge, London)

Bohm, D. (1980): *Wholeness and the Implicate Order* (Routledge, London)

Bohm, D. (1982): Response to Schindler's Critique of My *Wholeness and the Implicate Order*, *Int. Philos. Q.* **22**(4), 329–339

Bohm, D. (1985): *Unfolding Meaning: A Weekend of Dialogue with David Bohm*, ed. by D. Factor (Foundation House Publications, Mickleton, UK) [republished by Routledge, London, 1987]

Bohm, D. (1986): Time, the Implicate Order, and Pre-space, in *Physics and the Ultimate Significance of Time: Bohm, Prigogine, and Process Philosophy*, ed. by D. Griffin (State University of New York Press, Albany NY)

Bohm, D. (1987a): Hidden Variables and the Implicate Order, in *Quantum Implications: Essays in Honour of David Bohm*, ed. by B. J Hiley and F. D. Peat (Routledge, London)

Bohm, D. (1987b): Discussions, in *Symposium on the Foundations of Modern Physics 1987, Joensuu, Finland*

Bohm, D. (1987c): The Implicate Order and Prigogine's Notions of Irreversibility, *Found. Phys.* **17**(7), 667–677

Bohm, D. (1988): Postmodern Science and a Postmodern World, in *The Reenchantment of Science*, ed. by D. R. Griffin (State University of New York Press, Albany NY)

Bohm, D. (1989): Meaning and Information, in *The Search for Meaning*, ed. by P. Pylkkänen (Crucible, Wellingborough, UK; distributed by HarperCollins, London)

Bohm, D. (1990): A New Theory of the Relationship of Mind and Matter, *Philos. Psych.* **3**(2), 271–286

Bohm, D. (1995): *On Dialogue* (Routledge, London)

Bohm, D. & Biederman, C. (1999): *The Bohm–Biederman Correspondence. Volume 1: Creativity and Science*, ed. by P. Pylkkänen (Routledge, London)

Bohm, D. & Hiley, B. J. (1984): Generalization of the Twistor to Clifford Algebras as a Basis for Geometry, *Revista Brasileira de Fisica, Volume Especial Os 70 anos de Mário Schönberg*, pp. 1–26

Bohm, D. & Hiley, B. J. (1985): Active Interpretation of the Lorentz "Boosts" as a Physical Explanation of Different Time Rates, *Am. J. Phys.* **53**(8), pp. 720–723

Bohm, D. & Hiley, B. J. (1987): An Ontological Basis for Quantum Theory: I. Non-relativistic Particle Systems, *Phys. Rep.* **144**(6), 323–348

Bohm, D. & Hiley, B. J. (1993): *The Undivided Universe: An Ontological Interpretation of the Quantum Theory* (Routledge, London)

Bohm, D., Hiley, B. J. & Kaloyeroy, P. N. (1987): An Ontological Basis for the Quantum Theory II, *Phys. Rep.* **144**(32), 349

Bohm, D. & Peat, F. D. (2000): *Science, Order, and Creativity* 2nd edn (Routledge, London)

Bohm, D. & Pylkkänen, P. (1991): Cognition as a Movement toward Coherence (unpublished manuscript)

Broad, C. D. (1923): *Scientific Thought* (Kegan Paul, London)

Broad, C. D. (1925): *The Mind and Its Place in Nature* (Routledge and Kegan Paul, London)

Brown, H. R. & Harré, R. (eds) (1988): *Philosophical Foundations of Quantum Field Theory* (Clarendon Press, Oxford)

Bunge, M. (1979): *Causality and Modern Science*, 3rd revised edn (Dover, New York)

Callender, C. & Huggett, N. (eds) (2001): *Physics Meets Philosophy at the Planck Scale* (Cambridge University Press, Cambridge)

Carruthers, P. (2001): Higher-Order Theories of Consciousness, in *The Stanford Encyclopedia of Philosophy*, Summer 2001 edn, ed. by E. N. Zalta (http://plato.stanford.edu/archives/sum2001/entries/consciousness-higher)

Cartwright, N. (1983): *How the Laws of Physics Lie* (Clarendon Press, Oxford)

Chalmers, D. (1995): Facing Up to the Problem of Consciousness, *J. Consc. Stud.* **2**, 200–219

Chalmers, D. J. (1996): *The Conscious Mind: In Search of a Fundamental Theory* (Oxford University Press, Oxford)

Chalmers, D. J. (1997): Moving Forward on the Problem of Consciousness, *J. Consc. Stud.* **4**(1), 3–46

Chalmers, D. (2002a): Does Conceivability Entail Possibility, in *Conceivability and Possibility*, ed. by T. Gendler & J. Hawthorne (Oxford University Press, Oxford), pp. 145–200

Chalmers, D, (2002b): Consciousness and Its Place in Nature, in *Philosophy of Mind: Classical and Contemporary Readings*, ed. by D. Chalmers (Oxford University Press, Oxford) [also in *Blackwell Guide to the Philosophy of Mind*, ed. by S. Stich & F. Warfield (Blackwell, Oxford, 2003)]

Chrisley, R. C. (1997): Learning in Non-superpositional Quantum Neurocomputers, in *Brain, Mind and Physics*, ed. by P. Pylkkänen, P. Pylkkö & A. Hautamäki (IOS Press, Amsterdam), pp. 126–139

Churchland, P. M. (1984): *Matter and Consciousness: A Contemporary Introduction to the Philosophy of Mind* (MIT Press, Cambridge MA)

Churchland, P. M. (1989): *A Neurocomputational Perspective. The Nature of Mind and Structure of Science* (MIT Press, Cambridge MA)

Churchland, P. M. (2004): Functionalism at Forty: A Critical Retrospective, *J. Philos.* **CII**(1), 33–50

Churchland, P. S. (1986): *Neurophilosophy* (MIT Press, Cambridge MA)

Comfort, A. (1981): The Implications of an Implicate, *J. Social Biol. Struct.* **4**, 363–374

Cottingham, J. (1986): *Descartes* (Blackwell, Oxford)

Crane, T. (1992): Mental Causation and Mental Reality, *Proc. Aristotelian Soc.* **XCII**(2), 185–202

Crick, F. H. (1994): *The Astonishing Hypothesis: The Scientific Search for the Soul* (Scribner, New York NY)

Cushing, J. T. (1994): *Quantum Mechanics: Historical Contingency and the Copenhagen Hegemony* (University of Chicago Press, Chicago)

Cushing, J. T., Fine, A. & Goldstein, S. (eds) (1996): Bohmian Mechanics and Quantum Theory: An Appraisal, *Boston Studies in the Philosophy of Science* vol. 184 (Kluwer, Boston)

Dainton, B. (2000): *Stream of Consciousness. Unity and Continuity in Conscious Experience* (Routledge, London)

Dainton, B. (2001): *Time and Space* (Acumen, Chesham)

Dainton, B. (2004): Unity in the Void: Reply to Revonsuo, *Psyche* **10**(1) (http://psyche.cs.monash.edu.au/symposia/dainton/rev-r.pdf)

Davidson, D. (1970): Mental Events, in *Experience and Theory*, ed. by L. Foster & J. W. Swanson (University of Massachusetts Press, Amherst MA), pp. 79–101

Davidson, D. (2001): *Essays on Actions and Events* 2nd edn (Clarendon Press, Oxford)

Dehmelt, H. (1990): *Am. J. Phys.* **58** 17–27

Dennett, D. C. (1991): *Consciousness Explained* (Little, Brown, Boston MA)

Dennett, D. C. (1999): The Zombie Hunch: Extinction of an Intuition?, final draft for the Royal Institute of Philosophy Millennial Lecture, 28 November 1999 (http://www.nyu.edu/gsas/dept/philo/courses/consciousness/papers/DD-zombie.html)

Dretske, F. (1981): *Knowledge and the Flow of Information* (MIT Press, Cambridge MA)

Earman, J., Glymour, C. & Mitchell, S. (eds) (2002): *Ceteris Paribus Laws* (Kluwer, Dordrecht)

Faye, J. (1999) Explanation Explained, *Synthese* **120**, 61–75

Faye, J. (2002): Copenhagen Interpretation of Quantum Mechanics, in *The Stanford Encyclopedia of Philosophy*, Summer 2002 edn, ed. by E. N. Zalta (http://plato.stanford.edu/archives/sum2002/entries/qm-copenhagen/)

Feyerabend, P. K. (1960): Professor Bohm's Philosophy of Nature, *Brit. J. Philos. Sci.* **X**(40), 321–338

Feynman, R. P., Leighton, R. B. & Sands, M. (1965): *The Feynman Lectures on Physics. Vol. 3: Quantum Mechanics* (Addison Wesley, Reading MA)

Fodor, J. (1990): *A Theory of Content* (MIT Press, Boston MA)

Folse, H. (1985): *The Philosophy of Niels Bohr* (North-Holland, Amsterdam)

Forman, P. (1987): Behind Quantum Electronics: National Security as Basis for Physical Research in the United States, 1940–60, *Hist. Stud. Phys. Biol. Sci.* **18**, 149–229

Freire, O., Jr. (2005): Science and Exile: David Bohm, the Hot Times of the Cold War, and His Struggle for a New Interpretation of Quantum Mechanics, *Hist. Stud. Phys. Biol. Sci.* **36**(1), 1–34

Frescura, F. A. M. & Hiley, B. J. (1980): The Implicate Order, Algebras, and the Spinor, *Found. Phys.* **10**(1), 7–31

Ghirardi, G. (2002): Collapse Theories, in *The Stanford Encyclopedia of Philosophy*, Spring 2002 edn, ed. by E. N. Zalta (http://plato.stanford.edu/archives/spr2002/entries/qm-collapse/)

Gierer, A. (2002): Holistic Biology Back on stage? Comments on Post-genomics in Historical Perspective, *Philos. Nat.* **39**(1), 25–44

Gill, M. L. (2005): Aristotle's Metaphysics Reconsidered, *J. Hist. Philos.* **43**(3), 223–251

Globus, G. (1987): *Dream Life, Wake Life* (State University of New York Press, Albany NY)

Globus, G. (2003): *Quantum Closures and Disclosures* (John Benjamins, Amsterdam)

Globus, G., Pribram K. H. & Vitiello, G. (eds) (2004): *Brain and Being. At the Boundary between Science, Philosophy, Language, and Arts* (John Benjamins, Amsterdam)

Goldstein, S. (2002): Bohmian Mechanics, in *The Stanford Encyclopedia of Philosophy Philosophy*, Winter 2002 edn, ed. by E. N. Zalta (http://plato.stanford.edu/archives/win2002/entries/qm-bohm/)

Grush, R. & Churchland, P. S. (1995): Gaps in Penrose's Toilings, *J. Consc. Stud.*, **2**(1), 10–29

Guarini, M. (2003): Bohm's Metaphors, Causality, and the Quantum Potential, *Erkenntnis* **59**, 77–95

Güzeldere, G. (1997): *Introduction: The Many Faces of Consciousness: A Field Guide* in *The Nature of Consciousness: Philosophical Debates*, ed. by N. Block, O. J. Flanagan, & G. Güzeldere (MIT Press, Cambridge MA), pp. 1–67

Hackermüller, L., Hornberger, K., Brezger, B., Zeilinger, A. & Arndt, M. (2004): Decoherence of Matter Waves by Thermal Emission of Radiation, *Nature* **427**, 711–714

Hagan, S., Hameroff, S. R. & Tuszynzki, J. (2002): Quantum Computation in Brain Microtubules: Decoherence and Biological Feasibility, *Phys. Rev. E* **65**, 061901

Hameroff, S. R. & Penrose, R. (1996): Orchestrated Reduction of Quantum Coherence in Brain Microtubules: A Model for Consciousness?, in *Toward a Science of Consciousness The First Tucson Discussions and Debates*, ed. by S. R. Hameroff, A. W. Kaszniak & A. C. Scott (MIT Press, Cambridge MA)

Heidegger, M. (1962): *Being and Time* (Harper & Row, New York)

Heil, J. & Mele, A. (eds) (1995): *Mental Causation* (Oxford University Press, Oxford)

Hiley, B. J. (1980): Towards an Algebraic Description of Reality, *Annales de la Fondation Louis de Broglie* **5**(2), 75–103

Hiley, B. J. (1991): Vacuum or Holomovement, in *Philosophy of Vacuum*, ed. by S. Saunders & H. Brown (Clarendon Press, Oxford)

Hiley, B. J. (2004): Information, Quantum Theory, and the Brain, in *Brain and Being. At the Boundary between Science, Philosophy, Language, and Arts*, ed. by G. Globus, K. H. Pribram & G. Vitiello (John Benjamins, Amsterdam), pp. 197–214

Hiley, B. J. & Peat, F. D. (eds) (1987) *Quantum Implications: Essays in Honour of David Bohm* (Routledge, London)

Hiley, B. J. & Pylkkänen, P. (1996): Representation and Interpretation in Quantum Physics, in *Alternative Representations*, ed. by D. M. Peterson (Intellect Books, Exeter)

Hiley, B. J. & Pylkkänen, P. (1997) Active Information and Cognitive Science – A Reply to Kieseppä, in *Brain, Mind and Physics*, ed. by P. Pylkkänen, P. Pylkkö & A. Hautamäki (IOS Press, Amsterdam), pp. 64–85

Hiley, B. J. & Pylkkänen, P. (2001): Naturalizing the Mind in a Quantum Framework, in *Dimensions of Conscious Experience*, ed. by P. Pylkkänen & T. Vadén (John Benjamins, Amsterdam)

Hiley, B. J. and Pylkkänen, P. (2005) Can Mind Affect Matter via Active Information, *Mind and Matter* **3**(2), 7–27 (http://www.mindmatter.de/mmpdf/hileywww.pdf)

Holland, P. R. (1993): *The Quantum Theory of Motion* (Cambridge University Press, Cambridge)

Honderich, T. (ed) (1995): *The Oxford Companion to Philosophy* (Oxford University Press, Oxford)

Honderich, T. (2004): *On Consciousness* (Edinburgh University Press, Edinburgh)

Husserl, E. (1913): *Ideas: General Introduction to Pure Phenomenology*, translated by W. B. Gibson, 1931 edn (MacMillan, New York)

Isaacs, W. (1999): *Dialogue and the Art of Thinking Together* (Doubleday, New York)

Jantch, E. (1980): *The Self-organizing Universe* (Pergamon Press, Oxford)

Juti, R. (2001): *Johdatus Metafysikkaan* [*Introduction to Metaphysics*] (Gaudeamus, Helsinki) [in Finnish]

Kaila, E. (1956): Terminalkausalität als die Grundlage eines unitarischen Naturbegriffs. Eine naturphilosophische Untersuchung, *Acta Philos. Fenn.* **10**, 1–122

Kaila, E. (1979): *Reality and Experience*, ed. by R. S. Cohen (Reidel, Dordrecht)

Kant, I. (1991): *Critique of Pure Reason* 2nd edn (Dent, London) [originally published in German in 1787]

Kauffman, S. A. (1993): *The Origins of Order: Self-organization and Selection in Evolution* (Oxford University Press, Oxford)

Kauffman, S. A. (2000): *Investigations* (Oxford University Press, Oxford)

Kieseppä, I. A. (1997a) Is David Bohm's Notion of Active Information Useful in Cognitive Science?, in *Brain, Mind and Physics*, ed. by P. Pylkkänen, P. Pylkkö & A. Hautamäki (IOS Press, Amsterdam), pp. 54–63

Kieseppä, I. A. (1997b) On the Difference between Quantum and Classical Potentials – A Reply to Hiley and Pylkkänen, in *Brain, Mind and Physics*, ed. by P. Pylkkänen, P. Pylkkö & A. Hautamäki (IOS Press, Amsterdam), pp. 86–99

Kim, J. (1989): The Myth of Nonreductive Materialism, *Amer. Philos. Assoc. Proc.* **63**(3), 31–47

Kim, J. (1990): Explanatory Exclusion and the Problem of Mental Causation, in *Information, Semantics, and Epistemology*, ed. by E. Villanueva (Blackwell, Oxford)

Kim, J. & Sosa. E. (eds.) (1999): *Metaphysics: An Anthology* (Blackwell, Oxford)

Kohonen, T. (1984): *Self-organization and Associative Memory* (Springer, Berlin, Heidelberg)

Koskinen, H. (2004): *From a Metaphilosophical Point of View. A Study of W. V. Quine's Naturalism*, Acta Philos. Fennica LXXIV [The Philosophical Society of Finland, Helsinki]

Krishnamurti, J. & Bohm, D. (1999): *The Limits of Thought* (Routledge, London)

Laurikainen, K. V. (1988): *Beyond the Atom: The Philosophical Thought of Wolfgang Pauli* (Springer, Berlin, Heidelberg)

Le Poidevin, R. (2004): The Experience and Perception of Time, in *The Stanford Encyclopedia of Philosophy* Winter 2004 edn, ed. by E. N. Zalta (http://plato.stanford.edu/archives/win2004/entries/time-experience/)

Lehar, S. (2002) *The World in Your Head. A Gestalt View of the Mechanism of Conscious Experience* (Lawrence Erlbaum Associates, Mahwah NJ)

Lockwood, M. (1989): *Mind, Brain, and the Quantum* (Oxford University Press, Oxford)

Loux, M. J. (2002): *Metaphysics*, 2nd edn (Routledge, London)

Lycan, W. (1987): *Consciousness* (MIT Press, Cambridge MA)

Lycan, W. (2003): Philosophy of Mind, in *Blackwell Companion to Philosophy*, ed. by N. Bunnin & E. Tsui-James (Blackwell, Oxford)

Machamer, P., Darden, L. & Craver, C. (2000): Thinking about Mechanisms, *Philos. Sci.* **67**, 1–25

Maldacena, J. (2005): The Illusion of Gravity, *Scientific American*, November 2005

Marcer, P. & Mitchell, E. (2001): What is Consciousness? An Essay on the Relativistic Quantum Holographic Model of the Brain/Mind, Working by Phase Conjugate Adaptive Resonance, in *The Physical Nature of Con-*

sciousness, Advances in Consciousness Research, ed. by P. Van Loocke (John Benjamins, Amsterdam), pp. 145–174

Maroney, O. J. E. (2002): Information and Entropy in Quantum Theory, PhD Thesis, Department of Physics, Birkbeck College, University of London (http://arxiv.org/abs/quant-ph/0411172)

McCann, E. (1994): History: Philosophy of Mind in the Seventeenth and Eighteenth Centuries, in *A Companion to the Philosophy of Mind*, ed. by S. Guttenplan (Blackwell, Oxford)

Miller, R. B. (1990): Supervenience Is a Two-Way Street, *J. Philos* **87**, 695–701

Montero, B. (1999): The Body Problem, *Nous* **33**(3), 183–200

Murphy, T. S. (1998): Quantum Ontology, in *Deleuze and Guattari – New Mappings in Politics, Philosophy, and Culture*, ed. by E. Kaufman & K. J. Heller (University of Minnesota Press, Minneapolis)

Murphy, M. J. & O'Neill, L. A. J. *What Is Life?: The Next Fifty Years. Speculations on the Future of Biology* (Cambridge University Press, Cambridge)

Nagel, T. (1978): *Mortal Questions* (Cambridge University Press, Cambridge)

Nagel, T. (1995): Dualism, in *The Oxford Companion to Philosophy*, ed. by T. Honderich (Oxford University Press, Oxford), pp. 206–7

Nichol, L. (ed.) (2002): *The Essential David Bohm* (Routledge, London)

Nicolic, H. (2005): Bohmian Particle Trajectories in Relativistic Fermionic Quantum Field Theory, *Found. Phys. Lett.* **18**(2), 363–380

Niiniluoto, I. (1987): Varieties of Realism, in *Symposium on the Foundations of Modern Physics 1987*, ed. by P. Lahti & P. Mittelstaedt (World Scientific, Singapore)

Niiniluoto, I. (2000): *Critical Scientific Realism* (Oxford University Press, Oxford)

Noordhof, P. (1995): Panpsychism, in *The Oxford Companion to Philosophy*, ed. by T. Honderich (Oxford University Press, Oxford)

Norris, C. (2000): *Quantum Theory and the Flight from Realism: Philosophical Responses to Quantum Mechanics* (Routledge, London)

Olwell, R. (1999): Physical Isolation and Marginalization in Physics. David Bohm's Cold War Exile, *Isis* **90**, 738–756

Pagels, H. R. (1982): *The Cosmic Code: Quantum Physics as the Language of Nature* (Bantam, New York)

Passmore, J. (1966): *A Hundred Years of Philosophy* 2nd edn (Penguin, London)

Peat, F. D. (1996): *Infinite Potential. The Life and Times of David Bohm* (Addison Wesley, Boston MA)

Peat, F. D. & Briggs, J. (1986): *Looking Glass Universe* (Simon & Schuster, New York)

Penrose, R. (1989): *The Emperor's New Mind: Computers, Minds, and the Laws of Physics* (Oxford University Press, Oxford)

Penrose, R. (1994): *Shadows of the Mind* (Oxford University Press, Oxford)

Penrose, R. & Hameroff, S. R. (1995): What Gaps'? Reply to Grush and Churchland, *J. Consc. Stud.*, **2**(2), 99–112

Perus, M. (1995) Analogies of the Quantum and Neural Processing – Consequences for Cognitive Science, in *New Directions in Cognitive Science*, ed. by P. Pylkkänen & P. Pylkkö (Finnish Artificial Intelligence Society, Helsinki), pp. 115–123

Petitot, J., Varela, F. J., Pachoud, P. & Roy, J. M. (eds) (1999) *Naturalizing Phenomenology* (Stanford University Press, Stanford CA)

Philippidis, C., Dewdney, C. & Hiley, B. J. (1979): Quantum Interference and the Quantum Potential, *Il Nuovo Cimento*, **52** (1), 15–28

Pinch, T. J. (1977): What Does a Proof Do If It Does Not Prove. A Study of the Social Conditions and Metaphysical Divisions Leading to David Bohm and John von Neumann Failing to Communicate in Quantum Physics, in *The Social Production of Scientific Knowledge: Sociology of the Sciences*, vol. I, ed. by E. Mendelsohn, P. Weingart & R. Whitley (Reidel, Dordrecht)

Pines, D. (1987): The Collective Description of Particle Interactions: From Plasmas to the Helium Liquids, in *Quantum Implications: Essays in Honour of David Bohm*, ed. by B. J. Hiley & D. J. Peat (Routledge, London)

Plotnitsky, A. (1994): *Complementarity. Anti-epistemology after Bohr and Derrida* (Duke University Press, Durham NC)

Plotnitsky, A. (2002): *The Knowable and the Unknowable. Modern Science, Nonclassical Thought, and the Two Cultures* (University of Michigan Press, Ann Arbor MN)

Pribram, K. H. (1977): *Languages of the Brain: Experimental Paradoxes and Principles in Neuropsychology* (Wadsworth, Monterey CA)

Pribram, K. H. (1987) The Implicate Brain, in *Quantum Implications: Essays in Honour of David Bohm*, ed. by B. J. Hiley & D. J. Peat (Routledge, London)

Putnam, H. (1975): *Mathematics, Matter and Method. Philosophical Papers, vol. 1*, 2nd edn 1985 (Cambridge University Press, Cambridge)

Pylkkänen, P. (1985): Toward a Realist View of Information. Some Objectivist Concepts of Information, Meaning, and Knowledge. Masters Thesis, School of Mathematical and Physical Sciences, University of Sussex, UK

Pylkkänen, P. (1989a): Bohm, Plato, and the Dark Age of Cave Mechanics, in *The Search for Meaning*, ed. by P. Pylkkänen (Crucible, Wellingborough; distributed by HarperCollins, London)

Pylkkänen, P. (ed.) (1989b) *The Search for Meaning* (Crucible, Wellingborough; distributed by HarperCollins, London)

Pylkkänen, P. (1992): *Mind, Matter, and Active Information: The Relevance of David Bohm's Interpretation of Quantum Theory to Cognitive Science*, Reports from the Department of Philosophy, University of Helsinki, No. 2 (Department of Philosophy, University of Helsinki, Helsinki)

Pylkkänen, P. (1993): Some Methodological Remarks on Dennett's Multiple Drafts Theory of Consciousness, CTS Reports, 93-04 (Center for Theoretical Study, Charles University, Prague)

Pylkkänen, P. (1995): Mental Causation and Quantum Ontology, in *Mind and Cognition: Philosophical Perspectives on Cognitive Science and Artificial Intelligence, Acta Philos. Fennica Fasc.* **LVIII** (The Philosophical Society of Finland, Helsinki)

Pylkkänen, P. (1996a): On Baking a Conscious Cake with Quantum Yeast and Flour, in *Consciousness at a Crossroads of Cognitive Science and Phenomenology*, ed. by J. Shawe-Taylor (Imprint Academic, Thorverton)

Pylkkänen, P. (1996b) Voidaanko kvantti-ilmiöt ymmärtää prosessina? [Is It Possible to Understand Quantum Phenomena as a Process?], in *Tieto, totuus ja todellisuus* [*Knowledge, Truth, and Reality*], Prof. Ilkka Niiniluoto's Festschrift (Gaudeamus, Helsinki) [in Finnish]

Pylkkänen, P. (1999a): The Philosophical Implications of Quantum Level Active Information, in *Proceedings of CASYS '98, The Second International Confererence on Computing Anticipatory Systems*, ed. by Daniel Dubois, American Institute of Physics Conference Proceedings **465** (Springer, Berlin, Heidelberg)

Pylkkänen, P. (1999b): "Introduction" and "chapter summaries" in *The Bohm–Biederman Correspondence. Vol. 1: Creativity and Science* ed. by P. Pylkkänen (Routledge, London)

Pylkkänen, P. (2000): The Problem of Consciousness: Natural and Artificial, in *STep 2000 – Millenium of Artificial Intelligence* (Finnish Artificial Intelligence Society, Helsinki)

Pylkkänen, P. (2004a): Keeping it all together. Review of *Altered Egos: How the Brain Creates the Self* by T. Feinberg (Oxford University Press, Oxford, 2002), *Contemp. Psych. APA Rev. Books* **49**(1), 99–101

Pylkkänen, P. (2004b): Can Quantum Analogies Help Us to Understand the Process of Thought?, in *Brain and Being. At the Boundary between Science, Philosophy, Language, and Arts*, ed. by G. Globus, K. H. Pribram & G. Vitiello (John Benjamins, Amsterdam)

Pylkkänen, P. (2005): Reconceptualizing What We Know, Review of *On Consciousness*, by T. Honderich (Edinburgh University Press, Edinburgh), *J. Consc. Stud.* **12** (7), 93–95

Pylkkänen, P. (forthcoming): Does Dynamical Modeling Explain Time Consciousness?, in *Computing, Philosophy, and Cognitive Science* [working title], ed. by G. Dodig-Crnkovic, & S. Stuart (Cambridge Scholars Press, Cambridge)

Pylkkänen, P. & Pylkkö P. (eds) (1995): *New Directions in Cognitive Science* (Finnish Artificial Intelligence Society, Helsinki)

Pylkkänen, P. & Vadn, T. (eds) (2001): *Dimensions of Conscious Experience* (John Benjamins, Amsterdam)

Pylkkänen, P., Pylkkö P. & Hautamäki, A. (eds) (1997): *Brain, Mind, and Physics* (IOS Press, Amsterdam)

Pöppel, E. (1988): *Mindworks: Time and Conscious Experience* (Harcourt, Boston)

Quantum Universe Committee (2004): *Quantum Universe. The Revolution in 21st Century Particle Physics* (DOE/NSF High Energy Physics Panel, Pittsburgh PA) (http://www.interactions.org/quantumuniverse/qu/about/download.html)

Quine, W. V. O. (1960): *Word and Object* (Harvard University Press, Cambridge MA)

Rescher, N. (1996): *Process Metaphysics: An Introduction to Process Philosophy* (State University of New York Press, Albany NY)

Revonsuo, A. (1995): Consciousness, Dreams, and Virtual Realities, *Philos. Psych.* **8**, 35–58

Revonsuo, A. (2003) The Contents of Phenomenal Consciousness: One Relation to Rule Them All and in the Unity Bind Them [Commentary on *Stream of Consciousness. Unity and Continuity in Conscious Experience* by B. Dainton (Routledge, London) *Psyche* **9**(08) (http://psyche.cs.monash.edu.au/v9/psyche-9-08-revonsuo.html)

Revonsuo, A. (2005): *Inner Presence: Consciousness as a Biological Phenomenon* (MIT Press, Cambridge MA)

Robb, D. & Heil, J. (2005): Mental Causation *The Stanford Encyclopedia of Philosophy*, Spring 2005 edn, ed. by E. N. Zalta (http://plato.stanford.edu/archives/spr2005/entries/mental-causation/)

Rorty, R. (1979): *Philosophy and the Mirror of Nature* (Princeton University Press, Princeton NJ)

Ruhnau, E. (1995): Time-gestalt and the Observer, in *Conscious Experience*, ed. by T. Metzinger (Imprint Academic, Schöningh)

Sarfatti, J. (1997): How Thoughts Shape Matter (http://www.stardrive.org/bohm.html)

Saunders, S. (2002): Is the Zero-Point Energy Real?, in *Ontological Aspects of Quantum Field Theory*, ed. by M. Kuhlmann, H. Lyre & A. Wayne (World Scientific, Singapore)

Schaffer, J. (2003): Is There a Fundamental Level, *Nous* **37**(3), 498–517

Schiffer, S. (1991): Ceteris Paribus Laws, *Mind* **100**(397) 1–17

Schindler, D. L. (1982): David Bohm on Contemporary Physics and the Overcoming of Fragmentation, *Int. Philos. Q.* **22**

Seager, W. (1991): *The Metaphysics of Consciousness* (Routledge, London)

Searle, J. (1992): *The Rediscovery of the Mind* (MIT Press, Cambridge MA)

Senge, P. (1990): *The Fifth Discipline* (Doubleday, New York)

Shanon, B. (1991): Cognitive Psychology and Modern Physics: Some Analogies, *Eur. J. Cog. Psych* **3**(2), 201–234

Smith, Q. (2002) Why Cognitive Scientists Cannot Ignore Quantum Theory, in *Consciousness: New Philosophical Perspectives*, ed. by Q. Smith & A. Jolic (Oxford University Press, Oxford)

Steinhardt, P. J. & Turok, N. (2006): Why the Cosmological Constant is Small and Positive, *Science* 4 May 2006: 11262311 (http://www.sciencemag.org/cgi/content/abstract/1126231v1)

Stoljar, D. (2001): Physicalism, in *The Stanford Encyclopedia of Philosophy*, Spring 2001 edn, ed. by E. N. Zalta (http://plato.stanford.edu/archives/spr2001/entries/physicalism/)

Stonier, T. (1990): *Information and the Internal Structure of the Universe: An Exploration into Information Physics* (Springer, London)

Strawson, P. F. (1966): *The Bounds of Sense. An Essay on Kant's Critique of Pure Reason* (Routledge, London)

Sundqvist, F. (2003): *Perceptual Dynamics: Theoretical Foundations and Philosophical Implications of Gestalt Psychology, Acta Philos. Gothoburgensia* No. 16 (Department of Philosophy University of Gothenburg, Gothenburg)

Vaidman, L. (2002): Many-Worlds Interpretation of Quantum Mechanics, in *The Stanford Encyclopedia of Philosophy*, Summer 2002 edn, ed. by E. N. Zalta (http://plato.stanford.edu/archives/sum2002/entries/qm-manyworlds/)

Valentini, A. (2001): *Pilot-Wave Theory: An Alternative Approach to Modern Physics* (Cambridge University Press, Cambridge)

van Gelder, T. (1997): The Dynamical Alternative, in *The Future of the Cognitive Revolution*, ed. by D. M. Johnson & C. E. Erneling (Oxford University Press, Oxford), pp. 227–244

van Gelder, T. (1999): Wooden Iron? Husserlian Phenomenology Meets Cognitive Science, in *Naturalizing Phenomenology*, ed. by J. Petitot, F. J. Varela, P. Pachoud & Roy, J. M. (Stanford University Press, Stanford CA)

Van Gulick, R. (1995): What Would Count as Explaining Consciousness?, in *Conscious Experience*, ed. by T. Metzinger (Imprint Academic, Schöningh)

Van Gulick, R. (2004): Consciousness, in *The Stanford Encyclopedia of Philosophy* Fall 2004 edn, ed. by E. N. Zalta (http://plato.stanford.edu/archives/fall2004/entries/consciousness/)

Varela, F. J. (1999): The Specious Present: A Neurophenomenology of Time Consciousness, in *Naturalizing Phenomenology*, ed. by J. Petitot, F. J. Varela, P. Pachoud & J. M. Roy (Stanford University Press, Stanford CA)

Vitiello, G. (2001): *My Double Unveiled. The Dissipative Quantum Model of Brain* (John Benjamins, Amsterdam)

Weber, R. (ed.) (1986): *Dialogues with Scientists and Sages* (Routledge, London)

Weinstein, S. (2006): Quantum Gravity, in *The Stanford Encyclopedia of Philosophy*, Spring 2006 edn, ed. by E. N. Zalta (http://plato.stanford.edu/archives/spr2006/entries/quantum-gravity/)

Wittgenstein, L. (1974): *Tractatus Logico-Philosophicus* (Routledge, London)

Index

GPSR Compliance

The European Union's (EU) General Product Safety Regulation (GPSR) is a set of rules that requires consumer products to be safe and our obligations to ensure this.

If you have any concerns about our products, you can contact us on ProductSafety@springernature.com

In case Publisher is established outside the EU, the EU authorized representative is:

Springer Nature Customer Service Center GmbH
Europaplatz 3
69115 Heidelberg, Germany

Batch number: 10844669

Printed by Printforce, the Netherlands